Cleaner Combustion

Cleaner Combustion

Special Issue Editors

Derek Dunn-Rankin
Yu-Chien Chien

MDPI • Basel • Beijing • Wuhan • Barcelona • Belgrade

MDPI

Special Issue Editors
Derek Dunn-Rankin
University of California
USA

Yu-Chien Chien
University of California
USA

Editorial Office
MDPI
St. Alban-Anlage 66
4052 Basel, Switzerland

This is a reprint of articles from the Special Issue published online in the open access journal *Energies* (ISSN 1996-1073) from 2018 to 2019 (available at: https://www.mdpi.com/journal/energies/special_issues/Cleaner_Combustion)

For citation purposes, cite each article independently as indicated on the article page online and as indicated below:

LastName, A.A.; LastName, B.B.; LastName, C.C. Article Title. *Journal Name* **Year**, *Article Number*, Page Range.

ISBN 978-3-03921-477-8 (Pbk)
ISBN 978-3-03921-478-5 (PDF)

Cover image courtesy of Yu-Chien Chien.

Contents

About the Special Issue Editors

Derek Dunn-Rankin is a professor of Mechanical and Aerospace Engineering at the University of California, Irvine. He has 30 years of combustion science experience, with more than 100 peer-reviewed publications in this and related fields.

Yu-Chien Chien is the lead combustion scientist of the UCI Lasers, Flames, and Aerosols laboratory. In this role, she established the W.M. Keck Deep Ocean Power Science facility and conducted the first electric field effects on microgravity flames experiments on the International Space Station, and she guides a wide range of research associated with combustion power generation and controlling harmful emissions from them.

Preface to "Cleaner Combustion"

The Special Issue in Energies on "Cleaner Combustion" focuses on how the combination of fuel treatment, effective energy extraction, and emission mitigation in combustion can be optimized to reduce the environmental impact that threatens living systems because humans rely heavily on energy for basic necessities and economic development. We refer to "cleaner combustion" as an explicit acknowledgment that any improvements in this critical technology can have a significant global impact because of the ubiquitous nature of combustion. A growing population and growing standards of living have produced explosive growth in energy demand, and the environmental upset has started feeding back. This Special Issue includes a spectrum of research on cleaner combustion that helps to balance our needs for energy with cognizant respect for the environment. The goal of this collection is to identify the challenges and improvements of existing energy generation strategies in combustion, as well as reaching various pathways that resolve combustion emissions. The topics range from combustion and flame research that directly or indirectly impact the optimization of combustion processes using conventional carbon-based or hydrocarbon fuels, to options leading to a cleaner combustion outcome using alternative fuels, biofuels, or low carbon fuels, or including new methods to process exhausted gases.

Derek Dunn-Rankin, Yu-Chien Chien
Special Issue Editors

.

energies

MDPI

Article

Combustion Characteristics of Methane Hydrate Flames

Yu-Chien Chien * and Derek Dunn-Rankin *

Department of Mechanical and Aerospace Engineering, University of California, Irvine, CA 92697, USA
* Correspondence: chieny@uci.edu (Y.-C.C.); ddunnran@uci.edu (D.D.-R.)

Received: 15 April 2019; Accepted: 8 May 2019; Published: 21 May 2019

Abstract: This research studies the structure of flames that use laboratory-produced methane hydrates as fuel, specifically for the purpose of identifying their key combustion characteristics. Combustion of a methane hydrate involves multiple phase changes, as large quantities of solid clathrate transform into fuel gas, water vapor, and liquid water during burning. With its unique and stable fuel energy storage capability, studies in combustion are focused on the potential usage of hydrates as an alternative fuel source or on their fire safety. Considering methane hydrate as a conventional combustion energy resource and studying hydrate combustion using canonical experimental configurations or methodology are challenges. This paper presents methane hydrate flame geometries from the time they can be ignited through their extinguishment. Ignition and burning behavior depend on the hydrate initial temperature and whether the clathrates are chunks or monolithic shapes. These behaviors are the subject of this research. Physical properties that affect methane hydrate in burning can include packing density, clathrate fraction, and surface area. Each of these modifies the time or the temperature needed to ignite the hydrate flames as well as their subsequent burning rate, thus every effort is made to keep consistent samples. Visualization methods used in combustion help identify flame characteristics, including pure flame images that give reaction zone size and shape and hydrate flame spectra to identify important species. The results help describe links between hydrate fuel characteristics and their resulting flames.

Keywords: methane hydrate; gas hydrate; methane clathrate; hydrate combustion; hydrate flame spectrum; hydrate ignition; watery flames

1. Introduction

Gas clathrate hydrates are ice-like crystalline compounds with guest molecules caged by non-stoichiometric hydrogen-bonded water [1]. Burning gas hydrates has been seen and demonstrated as a very interesting phenomenon for future energy and environmental benefit. For example, there are significant methane reserves stored in the form of methane clathrates in sediment in the ocean's continental shelves and in permafrost. Permafrost stores of hydrate are particularly vulnerable to warming trends, and significant methane releases from these regions have been noted in the past [2,3]. The potential combustion-related technological issues include safety during gas storage, using hydrates as in situ thermal sources for additional hydrate dissolution, and clean power because hydrates represent a unique fuel that is remarkably diluted (considering the water content) but still flammable. Gas hydrate research in combustion science and the demonstration of its potential merit are relatively scarce [4–8], though hydrate studies in general remain active in chemistry as well as in efforts for natural hydrate resource exploration (e.g., [9–12]).

Combustion of methane hydrates involves multiple phase changes (not conventional for a fuel), which adds new controlling dimensions to the physics and the chemistry of the combustion problem, and some recent studies that were originally motivated by clean energy demand have begun to unravel some of the fundamental aspects of the direct combustion of methane hydrates. In some ways, the burning of a methane hydrate is similar to the devolatilization followed by volatiles combustion phases of coal burning [13]. That is, the first phase of coal combustion is the thermally-driven release of volatile hydrocarbons from the coal particle, which leaves behind the mineral and carbon-rich char. Similarly, thermally driven dissociation of hydrate releases volatile methane and leaves behind water. In both cases, the volatile combustible gas burns and produces the heat needed to continue the process. The difference is, of course, that the carbon-rich coal char is also combustible, whereas the residual water from hydrates is not. Nevertheless, the provision of energy to release sufficient volatiles that, when burning, contribute sufficient energy to maintain the subsequent release of flammable gas is the sequence necessary for continuous hydrate burning. Hence, in this paper, we describe the two important facets of the hydrate combustion process—ignition and steady burning of this unusual fuel.

Because the field is widely unexplored, even simple observations can contain important new information. For example, it can be seen from the hydrate combustion literature that the ignition for sustained burning must include a warming step to release sufficient methane gas to provide a robust flame capable of continued rapid hydrate dissociation. In addition, under nominally steady burning, it is observed from the literature (and from our experiments) that the hydrate flame color is different depending on the experiment. The current work provides a summary of gas hydrate burning, particularly focused on the ignition step and on flame visualization during hydrate combustion.

2. Materials and Methods

Artificial methane hydrate samples were created from ground ice solid phases with a 5.75 h heating cycle modified from Stern et al.'s standard hydrate formation procedure, operating around a peak pressure of 1500 psi [14]. The formation approach was to pressurize a sample of ice with methane gas and to then cycle the system across and through the hydrate equilibrium thermal boundary. Our ice-based hydrate samples were generally formed within a cylindrical mold that carried approximately 20 g of hydrate depending on the packing density for each operating condition. The production of reproducible (in terms of gas content and morphology) hydrate fuel samples is notoriously difficult, because hydrate formation depends on nucleation and growth behaviors at very small scales, thus extra care was taken to create reproducible samples. The process for making hydrates has already been extensively documented. Therefore, we only provided the skeleton of the process here. Nevertheless, it is important to recognize that even with carefully controlled conditions, hydrate internal structure can still vary in unpredictable ways. We therefore included sufficient repeat trials to ensure results with general applicability. The detailed procedure we used for producing the hydrate is documented in [15]. The hydrate formation literature has shown that the inclusion of sodium dodecyl sulfate (SDS), just at the ppm level, can promote hydrate growth rate, especially when forming hydrates from liquid [16–18]. We created ice-based hydrate samples with the addition of SDS in comparison to those with no SDS and compared them as they burned.

As mentioned in the introduction, the ignition process for hydrates is not trivial. This is because hydrates are thermodynamically unstable at room temperature and pressure, thus to stabilize the fuel long enough for it to be studied under burning conditions, the samples must first be chilled so that they do not spontaneously dissociate. This stabilizing procedure for us (and most others) involved quenching and storage in a liquid nitrogen cooled environment. Such procedures are standard in the hydrate research field [5,14], but the ultra-cold sample means that some warming is needed before the surface of the hydrate can release enough methane for ignition.

We conducted a series of hydrate ignition test experiments to understand this special solid fuel as it released cold flammable gases during dissociation in surroundings at room temperature and pressure. The first test allowed the cold hydrate sample temperature to rise on an aluminum foil base directly. One thermocouple was placed at the bottom of the aluminum foil and another was placed at 1 cm above the hydrate surface to monitor the temperature. The hydrate sample was ignited with a piezo ignitor and a butane flame lighter. The second test used an open-cup ignition apparatus and a procedure with a butane lighter as the combustion initiator. This method was a simplified version of the American Society for Testing and Materials (ASTM) Cleveland open cup test for measuring the flash point for fuels [19]. Because hydrates exhibit two phase transitions and three phases, two thermocouples were installed. One was located at the bottom of the cup monitoring the hydrate temperature, and another was installed 3.8 cm from the cup bottom and above the height of the hydrate sample (1.25–2.5 cm).

When ignition was being evaluated, the conditions of the hydrate and the ignitor proximity were carefully monitored and recorded. When burning studies were the goal, such as during the overall spectral scan of the hydrate flame, we simply used a butane lighter to bathe the hydrate in heat for its start. The hydrate flame spectrum was then probed using a Princeton Instruments SpectraPro 2300i with a PIXIS 400 detector. The spectrometer was calibrated with a Xenon lamp. The flame images were recorded with a standard digital single-lens reflex (DSLR) camera.

3. Results

3.1. Ignition

The hydrate ignition tests were conducted with partially-crushed samples from originally cylindrically-shaped samples (see Figure 1) to allow more methane gas to release by providing a larger surface area exposed to the ambient environment. The hydrate samples were generally around 80% clathration ratio of methane, as referenced from the absolute maximum based on the total ideal hydrate cavity potential. The ideal hydrate (which was not ever achieved in practice) has a 0.15 methane to ice mass ratio, as compared to our experimental samples with a ratio of approximately 0.12. All hydrate samples were wrapped with aluminum foil and cooled with liquid nitrogen until they were ready for the desired tests. The laboratory room temperature was 23.8 °C ± 0.5.

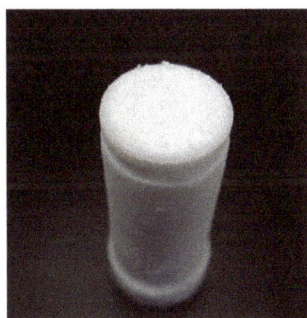

Figure 1. A typical sample of methane hydrate while in the Teflon cylindrical mold.

Figure 2 shows the top view (~50 cm²) and the side view of the initial ignition experiments. As is obvious, partially crushed hydrate samples varied in size and shape, but we found the behavior to be remarkably reproducible when the hydrate formation was reliable and the thermal conditions were controlled. The hydrate pile was approximately 3 cm tall. Both of the igniters were moving toward the hydrate sample with no physical contact during the process. The initial foil temperature was 2.3 °C, while the gas temperature was 22.5 °C. The hydrate samples could not be ignited with either the piezo igniter or the butane lighter in the first 60 s. The sample was too cold at this early time to release sufficient methane for flammability, though the methane inside the hydrate was dissociating with a distinct popping sound. At the 60 s time mark, the foil temperature read 16.4 °C, and the gas temperature was 23.7 °C. As the hydrate sample warmed up and the methane gas diffused slowly out of the sample, there was a distinct pattern of cold gas flowing down around the hydrate, as shown in Figure 3. The cold gases were flowing close to and along the foil, and they moved downward and away because of their relatively high density. The ignition by butane lighter of a steady flame occurred at the

89 s time mark, and at this time, the bottom side foil temperature was 18.5 °C. During ignition, the crushed powder hydrate region was the first spot lit into a methane flame, since the powder provided more methane volume and better air entrainment. Soon, a yellow methane flame was sustained out of the block shaped hydrate, including periodic rich methane jets, while a blue flame resided around the powder region with its more natural air admixing, as shown in Figure 4. Because the thermocouple located 10 mm above the sample was far away from the sample and measuring closed toward room temperature, it was measured at 23.0 °C at 2 mm above the hydrate sample at the point of ignition.

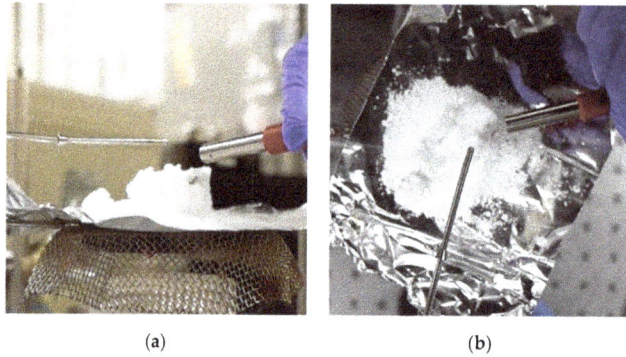

(a)   (b)

Figure 2. Methane hydrate ignition tests in open air environment, (**a**) side view (**b**) top view.

Figure 3. The methane hydrate warming up at room temperature. The arrows show the clear pathway of the cold gas flowing along the foil surface and moving downward. These flows were confirmed with both visual and schlieren imaging.

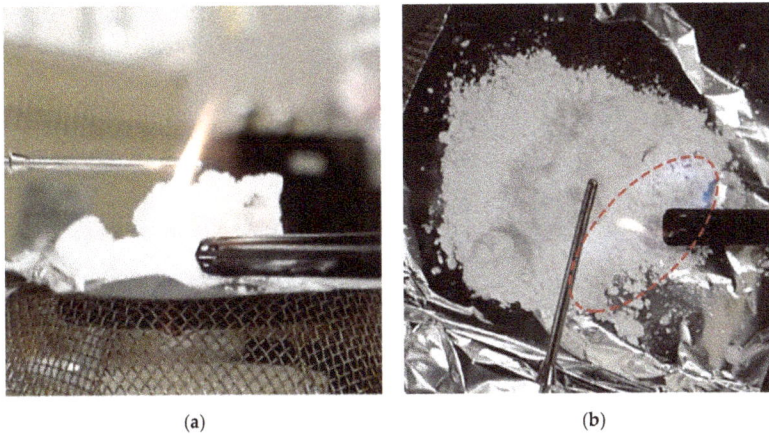

(a)　　　　　　　　　　　　　　　(b)

Figure 4. The snapshot at the moment when the methane hydrate was ignited by a butane lighter. The bright yellow flame was from a methane jet releasing from the hydrate block, shown in (**a**), while a pale blue flame was distributed around the hydrate powder with more air admixing into the methane released from this section of the sample in (**b**).

There were several uncertainties observed from the ignition tests of methane hydrate exposed to open air. For example, the thermocouple situated above the sample was not able to provide enough information, and the ignition location varied with the format and the geometry of the hydrate sample being evaluated. It was clear that a more standard test would be preferable in order to identify the hydrate ignition conditions, particularly with its special properties. The second test format was a cup burner modeled after the configuration designed to evaluate the ignition of other condensed fuels. In this more standard approach, the hydrate sample was placed inside a stainless-steel cup with two thermocouples monitoring temperatures, as shown in Figure 5. One was at the bottom of the cup, while the second one was 3.8 cm from the bottom. The hydrate pile was 1.3–2.5 cm tall (from powder chunks to large blocks). The butane lighter was used to observe any flash point or ignition that occurred during the process. The cup opening was half-covered throughout the process. The ignition or the flash point was observed when the upper thermocouple averaged −3.8 °C for two tests (−3.3 and −4.3 °C), while the lower thermocouple measured the hydrate samples as averaging in temperatures at −64.5 °C (−62.6 and −65.5 °C). A completely open cup without any cover was also used for comparison, and in this case, the measured upper temperature was −4.7 °C and the hydrate temperature was −36.8 °C.

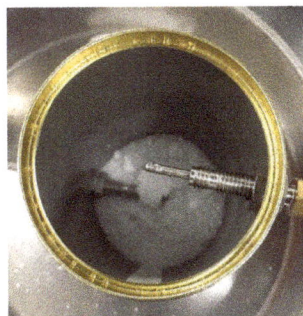

Figure 5. The cup ignition test with two thermocouples.

It should be noted that we recognize these ignition results are relatively sparse, but they are among the only attempts to quantify the characteristics of methane hydrate ignition. In addition, they reproducibly demonstrate the most important aspects of hydrate ignition that include a required minimum cake temperature, sufficient exposed surface area, and appropriate access to air for admixing. These elements are discussed more in later sections.

3.2. Burning

Figure 6 shows that, once burning, the flame had a distinct yellow-orange color when the samples included SDS. When SDS was not present, the flame had more of a blue-purple color around the gas hydrate with a stronger yellow color downstream. We presumed that the orange color arose from a sodium emission in the SDS case and that the yellow color in the post combustion zone could be nascent soot formation in the non-SDS hydrate. We also noted the foamy appearance in the case of the hydrate burning with SDS. We saw that excess amounts of SDS could make this foam a dominant feature that limited the hydrate combustion [6]. In order to compare the color and the character of the hydrate flame to one with similar characteristics of watery-fuel combustion but with the added feature that the flame be controllable (i.e., not affected by phase change processes), we developed a methane/air jet diffusion flame with varying amounts of water added to the methane. This gas-phase non-premixed flame simulated burning when water vapor naturally accompanied the fuel, as it did in a methane hydrate flame. Figure 7 shows a laminar methane diffusion jet flame as water was gradually added into the fuel to create a watery methane diffusion flame. The flame color changed from exhibiting a strong and brightly sooting region with no water addition to soot sitting on top of the flame tip as water was added. As even more water was added (up to the extinction limit), the flame color varied from blue to having a reddish hue. Based on these diffusion flame images, the formation of the blue-purple flame around the methane hydrate appeared related to the water vapor released into the flame. We expected that the water acted as a diluent to keep the flame temperature low and thereby avoid heavy sooting. To help confirm this feature, the gas hydrate optical spectrum was measured to identify the flame colors that appeared during combustion.

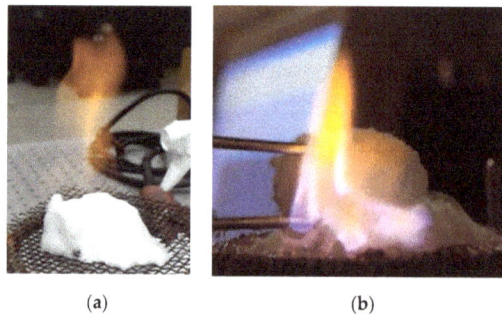

(a)   (b)

Figure 6. Gas hydrate burning (**a**) with surfactant sodium dodecyl sulfate (SDS) and (**b**) with no surfactant.

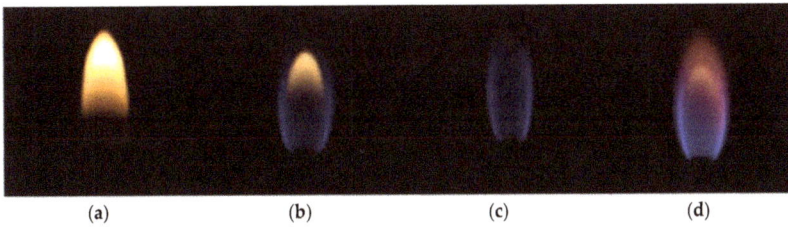

(a)   (b)   (c)   (d)

Figure 7. Methane/air laminar jet diffusion flame. Jet diameter was 2 mm, and methane flow rate was 65 mL/min with air coflowing 1 L/min. The figure shows the evolution of the flame with water addition; (**a**) no water added; (**b**) 0.25 water/methane molar ratio; (**c**) 0.55 water/methane molar ratio; (**d**) flame at the extinction limit (0.57 water/methane molar ratio).

In order to determine the flame spectral character, the probe of the spectrometer was focused toward where the hydrate flame started, very close to the methane hydrate cake (around 4 cm height). The spectrometer had a 100 msec frame rate, and the results were averaged over 600 frames with background and noise subtraction. The amplitude of the signal changed between frames, but the spectral shape was very stable. The initial spectrum, Figure 8, showed one major peak located near 600 nm and two minor peaks between 400–450 nm and 500–550 nm.

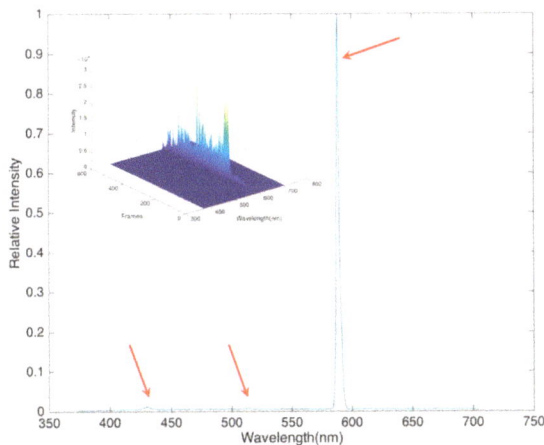

Figure 8. Gas hydrate spectrum over visible range; three peaks were found.

For a closer look, a different grating was used for the region close to 600 nm. With the higher resolution, two peaks were observed at 589.02 nm and 589.51 nm, as shown in Figure 9a. These two lines were clearly the sodium D line doublet (located at 589.0 and 589.6 nm). Over the 400–600 nm region, three peaks appeared—430.15 nm, 470.2 nm and 515.59 nm—which were the wavelength locations of the typical chemiluminescence species CH and C_2 in methane/air premixed flames [20–22], as shown in Figure 10. Perhaps unsurprisingly, the measured locations matched the three emission bands from the methane flames CH*(0,0), C_2*(1,0), and C_2*(0,0). Although these results were more confirmatory than exceptional, they did demonstrate that the spectral character of the flame was chemiluminescent and not strongly broadband, as would occur for soot. They also showed that the surfactants produced much of the strong color, and this was why the website images of methane hydrate combustion all had a strong orange-yellow coloring for dramatic presentation. A clean hydrate burning had much less luminosity and a pale blue color.

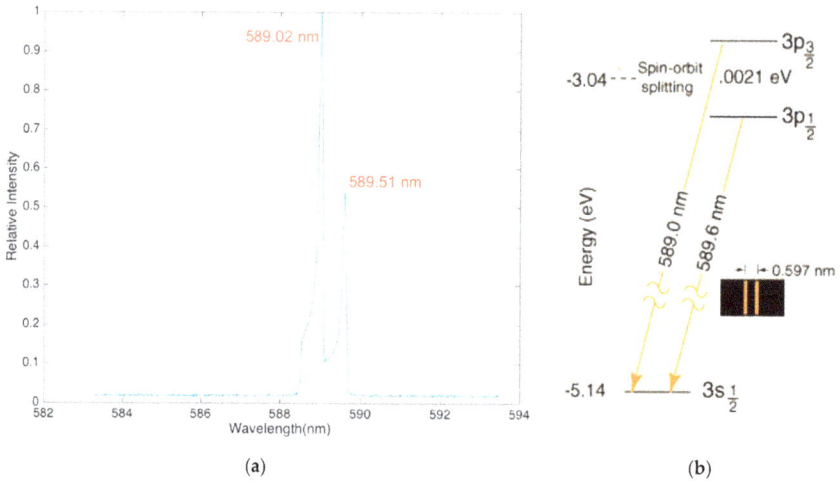

Figure 9. (a) Observed two peaks and (b) sodium D line doublet (courtesy of HyperPhysics from Georgia State University [23]).

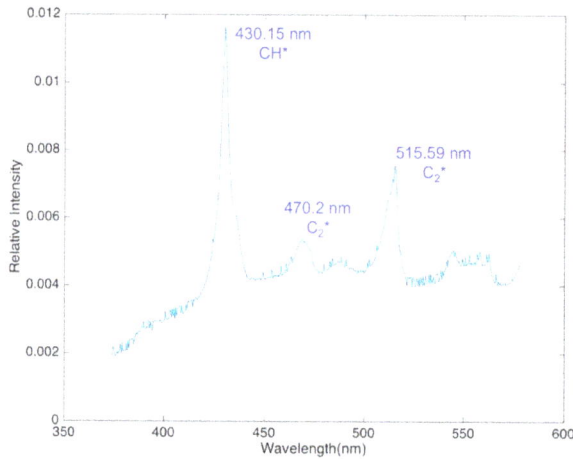

Figure 10. CH*(0,0), C2*(1,0), and C2*(0,0) emission bands of a methane hydrate flame.

4. Discussion

4.1. Ignition

Hydrate ignition is a complex process that is still not fully quantified. The results of this study show that stable burning requires that the sample reach some minimum temperature and that temperature can vary depending on the shape of the hydrate sample. Yoshioka et al. found that their cake temperature needed to reach −25 °C at the center for their spherical hydrates to burn steadily, and we found that temperatures of at least −4 °C are needed for steady burning of cylindrical cakes and powder piles. The reason for the minimum temperature is clear in that sufficient heat is needed from the flame to raise the hydrate temperature to dissociation at a rate sufficient to sustain the heat release needed for further dissociation. It seems that the cup burner is the most appropriate apparatus for determining hydrate ignition conditions.

4.2. Combustion

The spectral information during hydrate burning clearly shows the expected chemilumiscent emission and the influence of sodium from typical surfactants. The sodium line is also dominant in hydrates formed from salt water. We explored the use of sodium as a natural marker for temperature using a variant of the classic method of sodium line reversal [24]. Calibrating a tungsten lamp with varying input voltages against the blackbody emission measured from the filament provided a temperature/emission relationship. Then, using the lamp behind a hydrate flame, the voltage at which the lamp filament was just visible through the flame measured the hydrate flame's sodium temperature. The technique was promising but gave temperatures over 1900 K, which were higher than would be expected for a water diluted methane/air diffusion flame. The accurate measurement of hydrate flame temperature remains an experimental challenge we are pursuing with thin filament pyrometry (TFP). We measured the temperature of a methane flame fed by a melting hydrate [25], but thus far, there have been no temperature measurements of self-sustained natural hydrate flames.

In addition to the spectral information provided by the visual observation of the methane hydrate flames, there is also information contained in the time history of the total luminosity of the flame, as shown in the Figure 8 inset. The spectral measurements show that the proportion of the total flame luminosity arising from the different species remains nearly constant throughout the burn. That is, the flame color does not vary substantially in time. The total intensity, however, does vary as the flame fluctuates around the hydrate source. The buoyant flow driven by the rising hot gases also fluctuates, which gives a dynamic character to the flame. In the case of ideal spherical hydrate combustion [5], the flame fluctuation is clearly attributable to the growth of surface water that then drips from the base of the sphere, interrupting the flame anchor point and creating a fluctuation. In the case of our experiments, where the sample was more a distributed cake, the same effect of water could be observed, but it was not at a steady frequency, and thus the flame varied around the cake. A fast Fourier transform of the luminosity response showed one fairly dominant low frequency that arose from this buoyant plume shedding and water dripping frequency, and then a broader spectrum of frequency as the flame danced around the hydrate as methane was released.

5. Conclusions

This research examines the ignition and the burning of methane hydrates with particular attention to the repeatable tests for ignition and the evaluation of burning rates. Since this fuel is so unusual, the standard practices in combustion experiments required some modification and reformulation. Burning rates, for example, include the total mass loss from the sample but need to separate which part is water and which part is fuel. For ignition, a standard condition is needed for self-sustained combustion that is different than typical ignition definitions, where there is not such a large residual heat capacity in the inert matrix. Biofuels, particularly wood, have some similarities in this regard, and thus we used ignition and burning concepts from that literature in our study.

The ignition studies in the cup burner format show that the hydrate ignited at the temperature around −65 °C, while the gas temperature above was −4 °C for the half-covered tests. This measurement required a detailed refinement in procedures, such as the amount of the hydrate samples, the location of the thermocouples, and especially the location of the igniter, which was needed to further define. Therefore, standardizing a preparation and ignition method for methane hydrates is important. The literature describes preparation of hydrates in a variety of ways that leaves a residual chill in the hydrate cake. This chill (often from storage in liquid nitrogen) means that the hydrate cake is supercooled. In this case, when the liquid water forms as the hydrate dissociates, it is possible that the cold core then freezes this water and blocks further fuel release, thereby quenching the flame. In naturally dissociating hydrates, this process is called self-healing [14]. For burning, once the hydrate core has reached a temperature that allows it to dissociate continuously, self-healing does not occur, but identifying if that temperature is independent of hydrate morphology and shape remains to be done.

We presented measurements of the visible optical spectrum of direct methane hydrates in combustion. A spectrometric evaluation identified the species responsible for the observed flame color. The strong sodium doublet is responsible for most of the orange/yellow coloration attributed to hydrate flame images in the literature and online. The pale blue flame zone is much more subtle and, as expected, this region arises from the three chemiluminescence bands, namely $CH^*(0,0)$, $C_2^*(1,0)$, and $C_2^*(0,0)$. These baseline experiments provide the starting point for future study where we will examine the color of hydrate flames when the hydrates are formed with different surfactants (including biologically derived rhamnolipids) and where they may contain other contaminants. For example, we may explore the color of flames around hydrates formed from fracking flowback water. We recognize that the work shared here is mostly qualitative in nature, but as the study of methane hydrate combustion is still very new and considering its potential as a future energy source and an unusual format in combustion, we see this discussion as an interesting contribution to a cleaner combustion future.

Author Contributions: Y.-C.C. conceived, designed, and carried out the experiments. She also compiled and analyzed the resulting data. D.D.-R. shared in conceptualization, in the discussion and analysis of the results, and contributed to the hypotheses presented. Y.-C.C. wrote the paper with assistance from co-author D.D.-R.

Funding: This research was funded by W. M. Keck Foundation, with project title Carbon-Free Deep-Ocean Power Science.

Acknowledgments: The authors appreciate the support of the W. M. Keck Foundation for gas hydrate research in the University of California, Irvine—Deep Ocean Power Science Laboratory. The authors appreciate the enthusiastic participation in experiments from Adrien Ruas and David Escofet-Martin.

Conflicts of Interest: The authors declare no conflict of interest.

References

1. Sloan, E.D., Jr. *Clathrate Hydrates of Natural Gases: Revised and Expanded*, 2nd ed.; CRC Press: Boca Raton, FL, USA, 1998; ISBN 978-0-8247-9937-3.
2. The, U.S. Geological Survey Gas Hydrates Project Climate. Available online: https://woodshole.er.usgs.gov/project-pages/hydrates/database.html (accessed on 13 March 2019).
3. Yin, Z.; Linga, P. Methane hydrates: A future clean energy resource. *Chin. J. Chem. Eng.* **2019.** [CrossRef]
4. Roshandell, M.; Santacana-Vall, J.; Karnani, S.; Botimer, J.; Taborek, P.; Dunn-Rankin, D. Burning Ice—Direct Combustion of Methane Clathrates. *Combust. Sci. Technol.* **2016**, *188*, 2137–2148. [CrossRef]
5. Yoshioka, T.; Yamamoto, Y.; Yokomori, T.; Ohmura, R.; Ueda, T. Experimental study on combustion of a methane hydrate sphere. *Exp. Fluids* **2015**, *56*, 192. [CrossRef]
6. Wu, F.-H.; Chao, Y.-C. A Study of Methane Hydrate Combustion Phenomenon Using a Cylindrical Porous Burner. *Combust. Sci. Technol.* **2016**, *188*, 1983–2002. [CrossRef]
7. Nakoryakov, V.E.; Misyura, S.Y.; Elistratov, S.L.; Manakov, A.Y.; Sizikov, A.A. Methane combustion in hydrate systems: Water-methane and water-methane-isopropanol. *J. Eng. Thermophys.* **2013**, *22*, 169–173. [CrossRef]
8. Nakoryakov, V.E.; Misyura, S.Y.; Elistratov, S.L.; Manakov, A.Y.; Shubnikov, A.E. Combustion of methane hydrates. *J. Eng. Thermophys.* **2013**, *22*, 87–92. [CrossRef]
9. Boswell, R.; Collett, T.S. Current perspectives on gas hydrate resources. *Energy Environ. Sci.* **2011**, *4*, 1206–1215. [CrossRef]
10. Koh, C.A.; Sum, A.K.; Sloan, E.D. State of the art: Natural gas hydrates as a natural resource. *J. Nat. Gas Sci. Eng.* **2012**, *8*, 132–138. [CrossRef]
11. Fujii, T.; Suzuki, K.; Takayama, T.; Tamaki, M.; Komatsu, Y.; Konno, Y.; Yoneda, J.; Yamamoto, K.; Nagao, J. Geological setting and characterization of a methane hydrate reservoir distributed at the first offshore production test site on the Daini-Atsumi Knoll in the eastern Nankai Trough, Japan. *Mar. Pet. Geol.* **2015**, *66*, 310–322. [CrossRef]
12. Li, J.; Ye, J.; Qin, X.; Qiu, H.; Wu, N.; Lu, H.; Xie, W.; Lu, J.; Peng, F.; Xu, Z.; et al. The first offshore natural gas hydrate production test in South China Sea. *China Geol.* **2018**, *1*, 5–16. [CrossRef]
13. Smoot, L.D.; Hedman, P.O.; Smith, P.J. Pulverized-coal combustion research at Brigham Young University. *Prog. Energy Combust. Sci.* **1984**, *10*, 359–441. [CrossRef]

14. Stern, L.A.; Circone, S.; Kirby, S.H.; Durham, W.B. Anomalous Preservation of Pure Methane Hydrate at 1 atm. *J. Phys. Chem. B* **2001**, *105*, 1756–1762. [CrossRef]

15. Biasioli, A. Methane Hydrate Growth and Morphology with Implications for Combustion. Master's Thesis, University of California, Irvine, CA, USA, 2015.

16. Zhong, Y.; Rogers, R.E. Surfactant effects on gas hydrate formation. *Chem. Eng. Sci.* **2000**, *55*, 4175–4187. [CrossRef]

17. Ganji, H.; Manteghian, M.; Sadaghiani zadeh, K.; Omidkhah, M.R.; Rahimi Mofrad, H. Effect of different surfactants on methane hydrate formation rate, stability and storage capacity. *Fuel* **2007**, *86*, 434–441. [CrossRef]

18. Botimer, J.D.; Dunn-Rankin, D.; Taborek, P. Evidence for immobile transitional state of water in methane clathrate hydrates grown from surfactant solutions. *Chem. Eng. Sci.* **2016**, *142*, 89–96. [CrossRef]

19. *Standard Test Method for Flash and Fire Points by Cleveland Open Cup Tester*; ASTM International: West Conshohocken, PA, USA, 2013.

20. Kojima, J.; Ikeda, Y.; Nakajima, T. Spatially Resolved Measurement of OH*, CH*, and C2* Chemiluminescence in the Reaction Zone of Laminar Methane/air Premixed Flames. *Proc. Combust. Inst.* **2000**, *28*, 1757–1764. [CrossRef]

21. Walsh, K.T.; Long, M.B.; Tanoff, M.A.; Smooke, M.D. Experimental and computational study of CH, CH*, and OH* in an axisymmetric laminar diffusion flame. *Symp. (Int.) Combust.* **1998**, *27*, 615–623. [CrossRef]

22. Dandy, D.S.; Vosen, S.R. Numerical and Experimental Studies of Hydroxyl Radical Chemiluminescence in Methane-Air Flames. *Combust. Sci. Technol.* **1992**, *82*, 131–150. [CrossRef]

23. Thornton, S.T.; Rex, A. *Modern Physics for Scientists and Engineers*; Cengage Learning: Boston, MA, USA, 2012; ISBN 978-1-133-71223-7.

24. Padilla, R.E.; Minniti, M.; Jaimes, D.; Garman, J.; Dunn-Rankin, D.; Pham, T.K. Thin Filament Pyrometry for Temperature Measurements in Fuel Hydrate Flames and Non-Premixed Water-Laden Methane-Air Flames. In Proceedings of the 8th National Combustion Meeting, Park City, UT, USA, 19–22 May 2013.

25. Wu, F.H.; Padilla, R.E.; Dunn-Rankin, D.; Chen, G.B.; Chao, Y.C. Thermal structure of methane hydrate fueled flames. *Proc. Combust. Inst.* **2017**, *36*, 4391–4398. [CrossRef]

![energies logo] *energies*

MDPI

Article

Numerical Study of the Structure and NO Emission Characteristics of N_2- and CO_2-Diluted Tubular Diffusion Flames

Harshini Devathi *, Carl A. Hall and Robert W. Pitz

Department of Mechanical Engineering, Vanderbilt University, Nashville, TN 37235, USA;
carl.a.hall@vanderbilt.edu (C.A.H.); robert.w.pitz@vanderbilt.edu (R.W.P.)
* Correspondence: harshini.devathi@vanderbilt.edu

Received: 28 February 2019; Accepted: 13 April 2019; Published: 19 April 2019

Abstract: The structure of methane/air tubular diffusion flames with 65% fuel dilution by either CO_2 or N_2 is numerically investigated as a function of pressure. As pressure is increased, the reaction zone thickness reduces due to decrease in diffusivities with pressure. The flame with CO_2-diluted fuel exhibits much lower nitrogen radicals (N, NH, HCN, NCO) and lower temperature than its N_2-diluted counterpart. In addition to flame structure, NO emission characteristics are studied using analysis of reaction rates and quantitative reaction pathway diagrams (QRPDs). Four different routes, namely the thermal route, Fenimore prompt route, N_2O route, and NNH route, are examined and it is observed that the Fenimore prompt route is the most dominant for both CO_2- and N_2-diuted cases at all values of pressure followed by NNH route, thermal route, and N_2O route. This is due to low temperatures (below 1900 K) found in these highly diluted, stretched, and curved flames. Further, due to lower availability of N_2 and nitrogen bearing radicals for the CO_2-diluted cases, the reaction rates are orders of magnitude lower than their N_2-diluted counterparts. This results in lower NO production for the CO_2-diluted flame cases.

Keywords: tubular diffusion flame; methane/air; NO emissions; quantitative reaction pathway diagrams

---

1. Introduction

Laminar tubular flames are highly stretched and curved similar to turbulent flames and thus allow for isolating and studying stretch and curvature effects in flames [1]. In the past, tubular flames have been studied at 1 bar to understand the effect of stretch and curvature on preferential diffusion in hydrogen [2–12] and hydrocarbon flames [13] in premixed [2–5], non-premixed [6–11,13–16], and partially premixed configurations [12]. However, at high pressures, which are also characteristic of real-life combustors, minimal literature exists in the field of premixed tubular flames [17] and nothing in the field of non-premixed tubular diffusion flames. The authors in Nishioka et al. [17] numerically studied the structure of stoichiometric methane/air premixed tubular flames at different pressures. They observed that as the pressure was increased, the flame got thinner due to reduced thermal and species diffusivities. In addition, the peak temperature increased due to accelerated three-body reaction rates and approaches equilibrium values. In this paper, a detailed study of the structure of CH_4/air tubular diffusion flames where the fuel stream is diluted with either N_2 or CO_2 is reported.

In high-temperature combustion, which is characteristic of the modern-day jet engine and automotive combustors, pollutants such as NO are formed. When exhausted into the atmosphere, they lead to environmentally damaging effects such as acid rain, greenhouse effect, etc. [18]. Modern aero-propulsion and power generation gas turbines operate at high pressure (10–60 atm) and the forward thermal rate of NO formation (ppm/ms) increases dramatically with pressure [19].

Pressure effects on nitric oxide formation must be accurately understood to mitigate the adverse effects of pressure. A common practice of reducing NOx emissions is exhaust gas recirculation or air dilution. This is captured in part in this study using N_2 or CO_2 dilution of the fuel. In the past, NO emission characteristics have been studied for opposed jet diffusion flames. See, for example, [20–22]. The authors in Park et al. [20] studied the effect of preheat and CO_2 dilution of the oxidizer stream on NO emission rates in H_2/air counterflow diffusion flames. They observed that CO_2 suppresses the flame strength/temperature and thus also reduces NO emissions due to reduced thermal NO production. The researchers in Yang and Shih [23] studied NO formation in syngas flames. At low stretch rate (<10 s^{-1}) where radiation loss leads to low temperature, they found that $N + CO_2 \longleftrightarrow NO + CO$ was important contributor to NO formation in CO-rich zones. The authors of Lim et al. [21] also studied the effect of air preheating on the flame structure and NO emission characteristics of methane/air counterflow diffusion flames. They observed that with increase in temperature, the NO profiles showed a 70% increase which was attributed to the increased reaction rate of the prompt NO initiation reaction. Though NO production through the thermal route increased, it still remained an insignificant portion of the total NO production. The effect of fuel diluents, N_2, CO_2, and He, on NO profiles was studied by Rørtveit et al. [22] in H_2/air diffusion flames. It was observed that CO_2 and He dilution reduced the flame temperatures by large values and gave rise to lower NO emissions. However, the flames diluted with N_2 showed higher temperatures and higher NO mole fractions mainly coming through the thermal route. It was also noted that CO_2 reduced flame temperature largely through its dissociation. The researchers Shih and Hsu [24] have undertaken a numerical study to identify the important reaction pathways for NO production for hydrogen-lean and hydrogen-rich syngas flames. The thermal route was identified to be the most dominant and increased the NO production for hydrogen-rich syngas flames due to increased temperatures. Most of the above studies have used either quantitative reaction pathway diagrams, analysis of key reaction rates, or both to study the NO emission characteristics. In opposed non-premixed tubular flames, the flame temperature will differ from the opposed jet flame due to the effect of curvature and preferential diffusion [1,13,25]. This change of flame temperature with curvature will also affect NO production particularly through the thermally sensitive Zeldovich mechanism. In addition to the authors' knowledge, no tubular flame study has investigated NO emission characteristics using quantitative reaction pathway diagrams (QRPDs) or reaction rate analysis and this forms the main focus of the current work. QRPDs and reaction rate analysis are employed for identifying key reaction pathways for NO production in methane/air tubular diffusion flames. The rest of the paper is organized as follows. In Section 2, the burner geometry and the numerical methods employed are discussed. In Section 3, flame structure and NO emission pathways are discussed. This is followed by the conclusions for the current work.

2. Burner Geometry and Numerical Model

A schematic of the non-premixed tubular burner used in this study is shown in Figure 1. A non-premixed tubular flame is produced through two radially opposed impinging jets emanating from the inner and outer nozzles. The inner nozzle lies at the axis of the burner. The outer nozzle is concentric to the inner nozzle and at a higher radius. Fuel with diluent are flown through the inner nozzle while oxidizer is flown through the outer nozzle. The flame surface coincides with the cylindrical stagnation surface of the burner. The resulting flame is stretched and curved allowing for isolating and examining stretch and curvature effects. As shown in the right side of Figure 1, in the current study, the opposed tubular flame is concave ("negatively" curved) to the methane diluted fuel. For the diluted CH_4 mixture, the opposed tubular flame temperature will be reduced compared to an opposed jet flame or a positively curved flame [1,13]. As numerically studied in Figure 15 of Hu and Pitz [13] for opposed air vs. 30%N_2/70%CH_4 flames, all the stretched flame temperatures decrease with stretch rate and are lower than the adiabatic flame temperature (1995 K). The highest reduction of the flame temperature due to the stretch rate is for the negatively curved (concave toward the fuel) flame considered here. The inner nozzle is 6.35 mm in diameter, the outer nozzle is 24 mm in diameter, and the height of both the inner and the outer nozzles is 8 mm. More details on the opposed tubular burner simulated in this study can be found elsewhere [12].

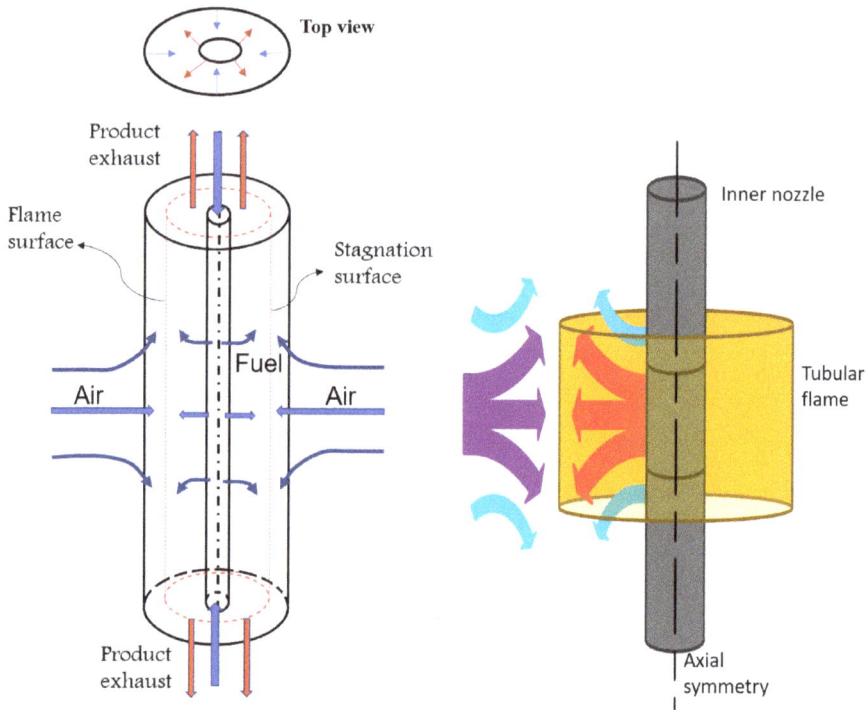

Figure 1. Schematic of non-premixed tubular burner.

Three-dimensional combustion equations can be reduced to two-dimensional form in r and θ using similarity transformation ([26], pp. 95–99) and further reduced to one-dimensional form in r with azimuthal symmetry [4,13]. The authors in Hall and Pitz [3] have developed a direct numerical simulation (DNS) code for two-dimensional tubular burner problems using the aforementioned similarity transformation and finite difference modeling. The code includes radiation heat loss from CO_2 and H_2O in the optically thin limit. The governing equations and details of the numerical model used in this work can be found in [3] and are not repeated here for the sake of brevity.

Analytical expressions for the cold global flow stretch rate have been derived in the past [27] for premixed and non-premixed tubular burner configurations. The stretch rate is defined as $k = 1/A \, dA/dt$ where A is the differential area of the flame [28]. For the non-premixed configuration, it can be expressed as

$$k = \pi \frac{V_i}{r_i} \left[\frac{(V_o r_o / V_i r_i) \sqrt{\rho_o / \rho_i} - 1}{r_o^2 / r_i^2 - 1} \right], \tag{1}$$

where V indicates radial velocity, r indicates radius, and ρ indicates mixture density. The subscripts i and o indicate inner and outer nozzle respectively.

In this study, methane/air diffusion flames are investigated at three different pressures: 1 atm, 3 atm, and 8 atm. The fuel stream is diluted by 65% with either N_2 or CO_2. The global stretch rate considered in this study is 88 s^{-1}. The GRI 3.0 mechanism [29] was used for chemical kinetic modeling. The mechanism, thermal, and transport data are processed using the CHEMKIN package.

Quantitative Reaction Pathway Diagrams

In addition to investigating the flame structure, this study also reports findings on the most important pathways that lead to NO emissions. Four different routes are investigated: Zelodovich or thermal route, prompt route, N_2O route, and NNH route. The important pathways are identified using reaction rate plots of primary NO formation reactions and quantitative reaction pathway diagrams (QRPDs). QRPDs are obtained by connecting the various species through arrows of varying thicknesses. The thickness of the arrows (or reaction paths) are proportional to the reaction rates of the various pathways integrated over the entire flame reaction zone [30,31].

3. Results

In this section, the structure of the tubular diffusion flame as a function of pressure and diluent is discussed. Following this, the various reactions that play a significant role in NO production are identified using QRPDs and sensitivity analysis.

3.1. Flame Structure

Figures 2 and 3 show the temperature, major and minor species profiles for N_2-diluted and CO_2-diluted CH_4/air flames at three different values of pressure: 1 atm, 3 atm, and 8 atm. Table 1 shows the boundary conditions used for the N_2- and CO_2-diluted cases at all pressures.

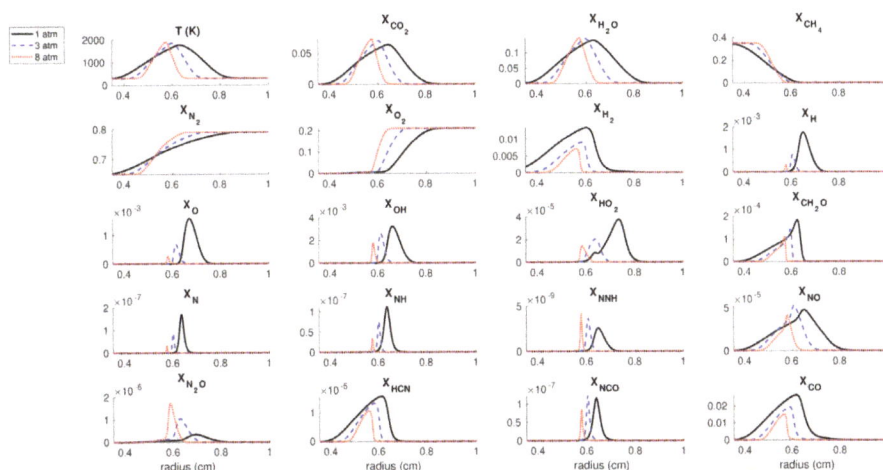

Figure 2. Temperature, major, and minor species profiles for 65% N_2-diluted CH_4/air tubular diffusion flame at 1 atm, 3 atm, and 8 atm; $k = 88$ s^{-1}.

Table 1. Boundary conditions for the simulated cases. V_o and W_o indicate radial velocity and axial velocity gradient at the outer air nozzle ($R = 1.2$ cm), V_i is the velocity at the inner fuel nozzle ($R = 0.32$ cm), k is the global stretch rate.

	$X_{CH4,i}$	$X_{N2,i}$	$X_{CO2,i}$	V_i (cm/s)	$X_{O2,o}$	$X_{N2,o}$	V_o (cm/s)	W_o (1/s)	k (1/s)
N_2-diluted case	0.35	0.65	0	15	0.21	0.79	25	15	88
CO_2-diluted case	0.35	0	0.65	15.09	0.21	0.79	30.24	20	88

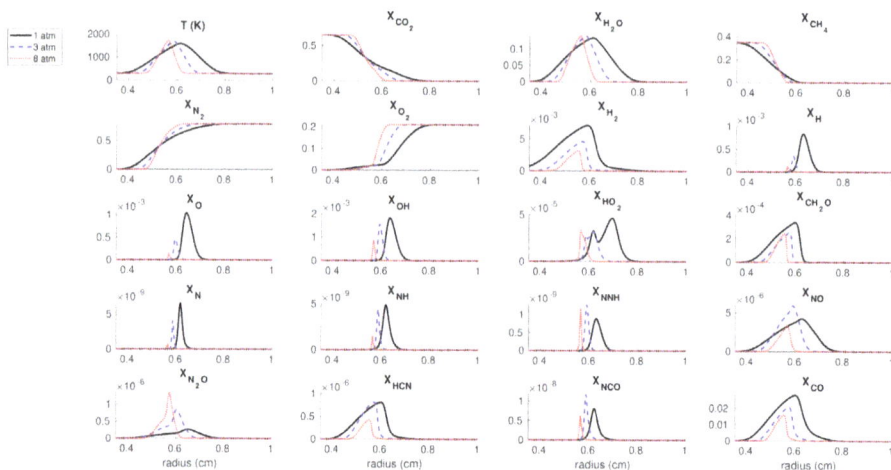

Figure 3. Temperature, major, and minor species profiles for 65% CO_2-diluted CH_4/air tubular diffusion flame at 1 atm, 3 atm, and 8 atm; $k = 88$ s^{-1}.

It is observed that the flame thickness reduces with pressure. This can be attributed to the reduced thermal diffusivity (α) with pressure (P) since $\alpha \sim 1/P$. In diffusion flames, $\delta_f \sim \sqrt{\alpha/k}$ [32], where δ_f is the flame thickness, α is the thermal diffusivity, and k is stretch rate. As the pressure is increased, the flame position as marked by temperature and species profiles (except for methane) is seen to move radially towards the nozzle. The peak values of temperature and major species remain mostly constant with pressure while that of the minor species mostly reduce with pressure. With pressure, reactions are more prone to go to completion due to reduced diffusivities (longer diffusion times). Minor/intermediate species concentrations reduce with pressure due to enhanced third-body collisions. The temperature of the flame for the CO_2-diluted case is lower than that for the N_2-diluted case. This is due to the fact that the specific heat of CO_2 is higher than that of N_2 and the radiative heat loss is higher for the CO_2-diluted flame. There is more heat transfer on the fuel side for the N_2-diluted flame than the CO_2-diluted flame. At 1 atm, the temperature reaches 300 K at 0.35 cm for the N_2-diluted flame and at 0.40 cm for the CO_2-diluted flame. The thermal diffusivity (α) is dependent on the molecular mass (m_i) according to $\alpha \sim m_i^{-1/2}$ leading to reduced value of α for CO_2 compared to N_2 dilution. NO emission is higher (10×) for N_2 diluted case than for CO_2-diluted case, as is to be expected due to availability of N_2 in abundance that leads to 20 times higher values of many nitrogen-bearing radicals. The temperature of the N_2-diluted flame is ~ 100 K higher as well. N_2O is formed from the third-body reaction $N_2 + O\,(^+M) \longleftrightarrow N_2O\,(^+M)$. Thus, its concentration increases with pressure. Concentrations of low-temperature radicals such as CH_2O and HO_2 are increased for the CO_2-diluted case as opposed to the N_2-diluted case due to reduced flame temperatures for the former. All N_2-based compounds show higher concentrations for N_2-diluted cases as is to be expected. For the CO_2-diluted flames, the absence of N_2 in the fuel leads to a change in slope for the N_2O profile as N_2O is formed by the reaction $N_2 + O + M \longleftrightarrow N_2O + M$. NO concentration increases with pressure and then decreases. This is discussed more in the upcoming sections where the various NO formation mechanisms are discussed using QRPDs and key reaction rate diagrams.

A comparison between the adiabatic flame temperature and the maximum flame temperature obtained using the DNS code mentioned in Section 2 for fuel dilution with CO_2 and N_2 at different values of pressure is shown in Table 2. The researchers in Hu and Pitz [13] have studied the effect of curvature on the temperature of CH_4/air flames with the fuel stream diluted by N_2. It was observed that the methane flame which is convex to the fuel stream shows a decrease in temperature from that of the planar stretched flames for certain values of global stretch rate. In the stretched 30%CH_4/70%N_2 vs. air flames, all the calculated flame temperatures were below the adiabatic flame temperature (1995

K). A similar behaviour can be observed in the current study as evident from Table 2. Due to the convex orientation of the flame front towards the fuel stream, the temperature of the tubular flame for all the cases shown in Figures 2 and 3 and Table 2 is lower than the adiabatic flame temperature. Radiation heat loss from CO_2 and H_2O reduces the flame temperature with the CO_2-diluted flame suffering the greater loss. At this time, there are no experimental measurements of tubular flames at high pressure. Experiments at high pressure imaging these non-premixed flames using the opposed tubular burner [12] are planned in the near future.

Table 2. Peak temperature obtained using the direct numerical simulation (DNS) code versus the adiabatic flame temperature for the N_2- and CO_2-diluted cases at different values of pressure.

Diluent	Pressure (atm)	T_{AD} (K)	$T_{peak,DNS}$ (K)	$\Delta T/T_{AD}$ (%)
N_2	1	2040	1770	-13.2
N_2	3	2050	1851	-9.7
N_2	8	2057	1898	-7.7
CO_2	1	1923	1598	-16.9
CO_2	3	1930	1692	-13.3
CO_2	8	1936	1719	-11.2

3.2. Quantitative Reaction Pathway Diagrams (QRPDs)

One of the main focuses of this study is also to analyze and identify the most important routes through which NO is produced. If NO production is not controlled, it can lead to a host of environmental problems such as acid rain, global warming, etc. A major step towards NO reduction is to identify the important routes to NO formation so that combustion chambers can be designed to target these specific routes. For example, if the thermal route contributes to the major portion of NO produced in a system, exhaust gas recirculation may be used to reduce the combustion chamber temperatures, thereby reducing NO production. Since the GRI 3.0 mechanism consists of about 350 reactions and analyzing all of them is beyond the scope of the current study, the most important reactions for NO production considered are as given in (Glassman et al. [18], pp. 403–414). These reactions are repeated in Table 3. QRPDs are a graphical representation of the reaction pathways between various species. Chemical compounds are connected with arrows indicating the direction of reaction flow. The size of the arrow is scaled proportional to the reaction rates that are integrated over the entire flame zone. Integrated reaction rates are obtained by integrating the individual reaction rates over the flame radius.

Figure 4 shows the quantitative reaction pathway diagrams for the N_2-diluted cases for $k = 88$ s^{-1} at three different values of pressure—1 atm, 3 atm, and 8 atm. Figure 4a shows the QRPD for 1 atm. The initiation step for the NNH route, $N_2 + H + M \longleftrightarrow NNH + M$, is the most dominant step. However, the subsequent chain propagation step, $NNH + O \longleftrightarrow NO + NH$, is less dominant than that leading to NO production through the Fenimore prompt route. The integrated reaction rates for all the steps in the Fenimore prompt route are, at the least, an order of magnitude higher than the remaining steps (except the initiation step for NNH route). However, the steps resulting in NO formation, $N + O_2 \longleftrightarrow NO + O$ and $N + OH \longleftrightarrow NO + H$, are common to both the thermal route and the Fenimore prompt route. Nevertheless, the N radical produced from the thermal route through the reaction $N_2 + O \longleftrightarrow NO + N$ is, at least, an order of magnitude less than the radical produced through the Fenimore prompt route via the reactions $NH + H \longleftrightarrow N + H_2$ and $NH + OH \longleftrightarrow N + H_2O$. This leads us to believe that the Fenimore prompt route is the most dominant route through which NO is produced. On the other hand, the N_2O route is much less significant and is several orders of magnitude less than the other reactions considered.

Table 3. Various NO formation routes considered in the current study and as taken from (Glassman et al. [18], pp. 403–414).

Route	Reaction
Thermal	$N + NO \longleftrightarrow N_2 + O$
	$N + O_2 \longleftrightarrow NO + O$
	$N + OH \longleftrightarrow NO + H$
Fenimore prompt	$HCN + O \longleftrightarrow NCO + H$
	$NCO + H \longleftrightarrow NH + CO$
	$NH + H \longleftrightarrow N + H_2$
	$NH + OH \longleftrightarrow N + H_2O$
	$N + O_2 \longleftrightarrow NO + O$
	$N + OH \longleftrightarrow NO + H$
N_2O	$N_2 + O(^+M) \longleftrightarrow N_2O(^+M)$
	$N_2O + O \longleftrightarrow 2\,NO$
NNH	$N_2 + H + M \longleftrightarrow NNH + M$
	$NNH + O \longleftrightarrow NO + NH$

As the pressure is increased to 3 atm (Figure 4b), the reaction rates for all the bimolecular reactions increase slightly but remain at the same order of magnitude. However, the third body reactions show order(s) of magnitude increases. Despite this, the N_2O route remains insignificant compared to the other routes. Due to the increase in reaction rates, the NO mole fraction as seen from Figure 2 also increases. With further increase in pressure to 8 atm, there is only a slight increase in the bimolecular reaction rates. As a matter of fact, for some of the reactions, the reaction rates decrease such as $N + OH \longleftrightarrow NO + H$ and $NH + H \longleftrightarrow N + H_2$ (Fenimore prompt). This could explain the reduction in NO for 8 atm. Further, it may be noted that the third body reaction rates do not see a significant increase from 3 atm to 8 atm as was seen from 1 atm to 3 atm. Again, N_2O route remains insignificant while the Fenimore prompt route is the most significant followed by NNH route and thermal route.

Figure 5 shows the QRPD for the CO_2-diluted cases for three different values of pressure as before. The reaction rates of all the reactions are order(s) of magnitude less that their N_2-diluted counterpart. This can be a result of the splitting up of CO_2 into CO and O as found by Rørtveit et al. [22] and further chemical inhibition by CO. It could also be result of decreased availability of N_2 for forming nitrogen compounds such as N that can lead to NO production. When comparing minor species profiles in Figures 2 and 3, the CO_2-diluted flame has NO producing radicals (N, NH, HCN, NCO) reduced by $20\times$. This is shown in Table 4. In addition, the reduced flame temperatures for the CO_2-diluted cases could also play a role. As the pressure is increased from 1 atm to 3 atm, the reaction rates increase slightly. However, similar to the N_2-diluted cases, the three-body reactions show order(s) of magnitude increase. Since CO_2 is a more efficient collision partner than N_2, as the pressure is increased from 3 atm to 8 atm, the reaction rate for $N_2 + H + M \longleftrightarrow NNH + M$ increases by an order of magnitude whereas it only increased slightly for the N_2-diluted case. Further, the reaction rates for all the bimolecular reactions decrease as the pressure is increased from 3 atm to 8 atm due to reduced diffusivities and molecular mixing as pressure is increased. Thus, the amount of NO produced is decreased more drastically from 3 atm to 8 atm as opposed to the N_2-diluted case. It may be noted that Fenimore prompt and NNH routes are still the most dominant followed by the thermal route whereas the N_2O route still remains insignificant.

(a)

(b)

(c)

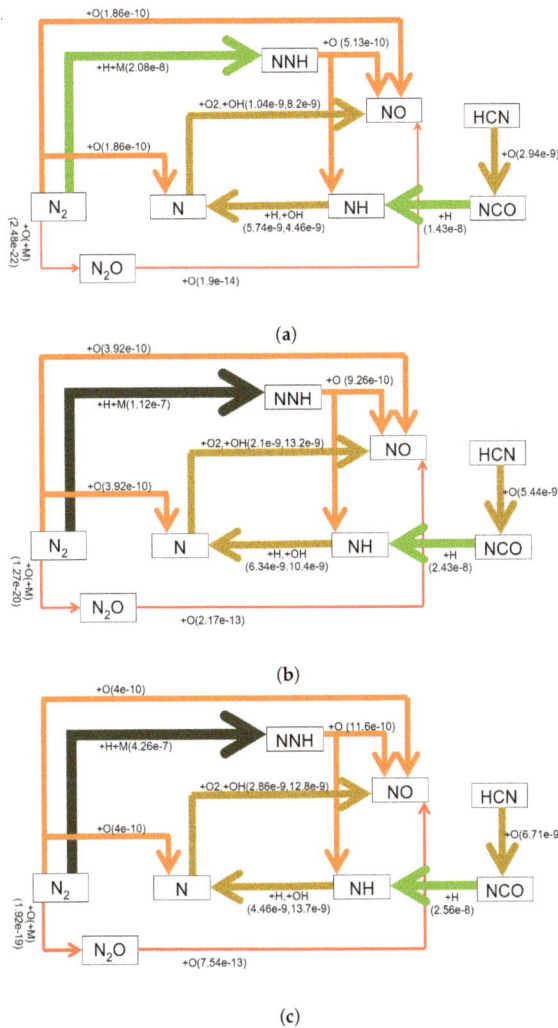

Figure 4. Quantitative reaction pathway diagrams (QRPDs) for the N_2-diluted case for $k = 88$ s^{-1} at (a) 1 atm, (b) 3 atm, and (c) 8 atm. For legends, please refer to Figure 6.

Table 4. Effect of CO_2 vs. N_2 dilution on maximum molefractions of NO producing radicals and NO. N, NH, HCN, NCO mole fractions are reduced by a factor of ∼20 for the CO_2-diluted case.

	1 atm			3 atm			8 atm		
	N_2 Case	CO_2 Case	Reduce by	N_2 Case	CO_2 Case	Reduce by	N_2 Case	CO_2 Case	Reduce by
N	2×10^{-7}	6×10^{-9}	33	8×10^{-8}	4×10^{-9}	20	2×10^{-8}	1×10^{-9}	20
NH	1×10^{-7}	5×10^{-9}	20	7×10^{-8}	4×10^{-9}	17	3×10^{-8}	2×10^{-9}	15
NNH	2×10^{-9}	8×10^{-10}	2.5	3×10^{-9}	1.3×10^{-9}	2.3	4×10^{-9}	1.2×10^{-9}	3.3
NO	4×10^{-5}	4×10^{-6}	10	5×10^{-5}	6×10^{-6}	8.3	3×10^{-5}	3×10^{-6}	10
N_2O	2×10^{-7}	2×10^{-7}	1	1×10^{-6}	6×10^{-7}	1.7	1.8×10^{-6}	1.2×10^{-6}	1.5
HCN	1.5×10^{-5}	7×10^{-7}	21	1.2×10^{-5}	7×10^{-7}	17	1×10^{-5}	3×10^{-7}	33
NCO	1.1×10^{-7}	7×10^{-9}	16	1.2×10^{-7}	1.1×10^{-8}	11	8×10^{-8}	6×10^{-9}	13

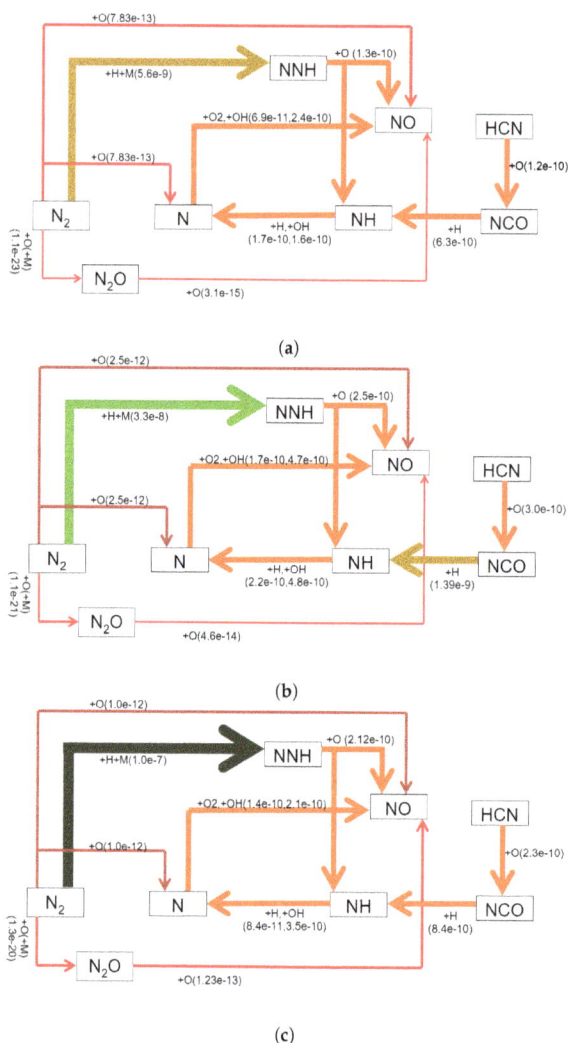

Figure 5. QRPD for the CO$_2$-diluted case for $k = 88$ s^{-1} at (**a**) 1 atm, (**b**) 3 atm, and (**c**) 8 atm. For legends, please refer to Figure 6.

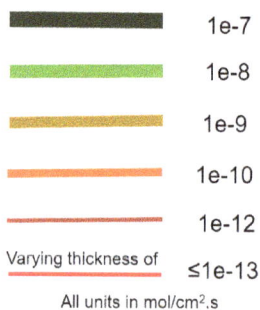

Figure 6. Legend for Figures 4 and 5.

3.3. Analysis of Reaction Rates

Reaction rates of various NO producing steps given in Table 3 are shown in Figure 7 as a function of radius for 1 atm and a stretch rate of 88 s^{-1}. Figures 8 and 9 show the same for 3 atm and 8 atm respectively. Note that for the CO_2-diluted cases, the reaction rates are multiplied by 10. It may be noticed that the radius range over which the reaction rates are non-zero decreases with pressure. This is to be expected due to the thinning of the flame zone due to reduced diffusivities. For a specific reaction, the reaction rate increases with pressure. As already indicated, Fenimore prompt appears to be the most dominant route to NO formation (Subfigures (b), (c), (f), and (g) of Figures 7–9) followed by the NNH route (Subfigure (e)), thermal route (Subfigure (a)), and N_2O route (Subfigure (d)). Further, it may be noticed that the ratio of reaction rates for the N_2-diluted case to CO_2-diluted case is approximately 1:10 except for the reactions $N_2O + O \longleftrightarrow 2\,NO$ and $NNH + O \longleftrightarrow NO + NH$. Both of these directly follow the third-body reactions as given under N_2O route and NNH route respectively in Table 3. Since CO_2 is an efficient third-body collision partner, the ratio of radicals N_2O and NNH (as shown in Figures 2 and 3) that are generated for the CO_2- and N_2-diluted cases is higher than the other radicals. This ready availability of these radicals, in turn, results in the increased reaction rates for $N_2O + O \longleftrightarrow 2\,NO$ and $NNH + O \longleftrightarrow NO + NH$ for the CO_2-diluted case when compared with the other reactions.

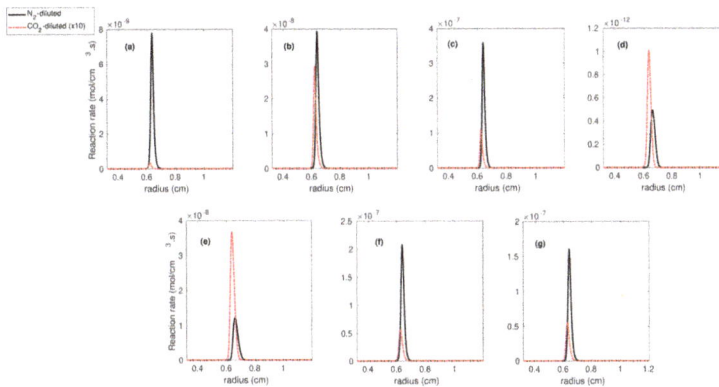

Figure 7. Reaction rate profiles at 1 atm as a function of radius for (**a**) $N_2 + O \longleftrightarrow N + NO$, (**b**) $N + O_2 \longleftrightarrow NO + O$, (**c**) $N + OH \longleftrightarrow NO + H$, (**d**) $N_2O + O \longleftrightarrow 2\,NO$, (**e**) $NNH + O \longleftrightarrow NO + NH$, (**f**) $NH + H \longleftrightarrow N + H_2$, and (**g**) $NH + OH \longleftrightarrow N + H_2O$.

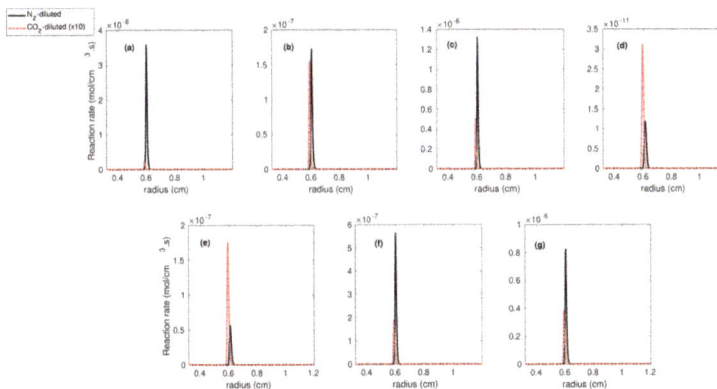

Figure 8. Reaction rate profiles at 3 atm as a function of radius for (**a**) $N_2 + O \longleftrightarrow N + NO$, (**b**) $N + O_2 \longleftrightarrow NO + O$, (**c**) $N + OH \longleftrightarrow NO + H$, (**d**) $N_2O + O \longleftrightarrow 2\,NO$, (**e**) $NNH + O \longleftrightarrow NO + NH$, (**f**) $NH + H \longleftrightarrow N + H_2$, and (**g**) $NH + OH \longleftrightarrow N + H_2O$.

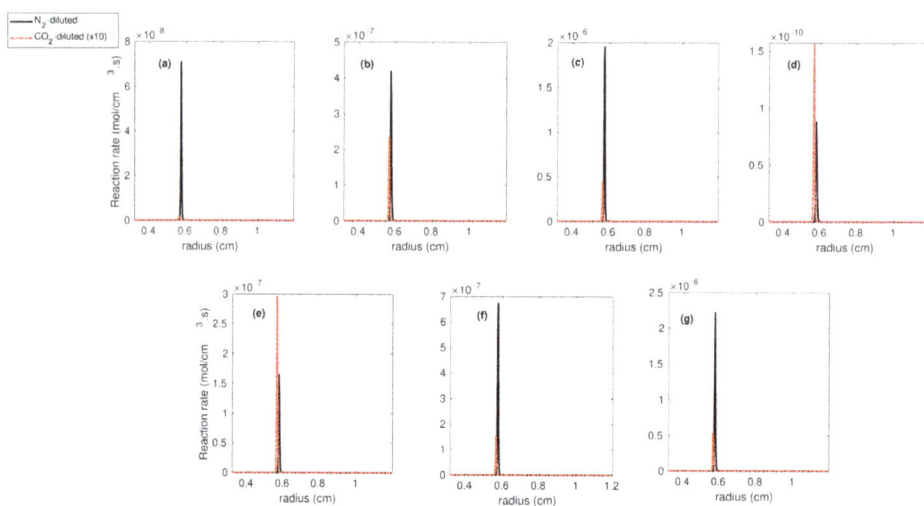

Figure 9. Reaction rate profiles at 8 atm as a function of radius for (**a**) $N_2 + O \longleftrightarrow N + NO$, (**b**) $N + O_2 \longleftrightarrow NO + O$, (**c**) $N + OH \longleftrightarrow NO + H$, (**d**) $N_2O + O \longleftrightarrow 2 NO$, (**e**) $NNH + O \longleftrightarrow NO + NH$, (**f**) $NH + H \longleftrightarrow N + H_2$, and (**g**) $NH + OH \longleftrightarrow N + H_2O$.

4. Conclusions

This paper has investigated the effect of pressure and 65% fuel dilution on the structure of non-premixed methane/air tubular diffusion flames. It was observed that due to reduced species and thermal diffusivities, the flame thickness decreases with pressure. The flame temperature for CO_2-diluted cases is lower than that for the N_2-diluted cases. This can be attributed to higher value of specific heat and higher radiation loss for CO_2 compared to N_2. Species formed through third-body collisions show increased concentrations with pressure. When N_2 dilution of the fuel is replaced by CO_2, the NO-producing radicals (N, NH, HCN, NCO) reduce by a factor of 20.

Quantitative reaction pathway diagrams (QRPDs) were obtained alongside performing an analysis of key reaction rates to identify the most important pathways for NO production. Four different NO production routes were considered, namely the thermal route, Fenimore prompt route, NNH route and the N_2O route. For both the CO_2- and N_2- diluted cases, it was found that the Fenimore prompt route is the most dominant followed by the NNH route, thermal route, and the N_2O route for all values of pressure. Though the integrated reaction rates for the N_2O route showed orders of magnitude increase with pressure, it remained an insignificant contributor to the overall NO production. An analysis of key reaction rates over the flame domain confirms the dominance of Fenimore prompt route in the production of NO. The NO production reaction rates for the CO_2-diluted cases were observed to be $10\times$ lower than for the N_2-diluted cases for most of the reactions. This was attributed to the greater availability of N_2 for the N_2-diluted cases, leading to $20\times$ higher peak values of many NO producing radicals.

Author Contributions: Conceptualization, H.D.; Methodology, H.D. and C.A.H.; Software, C.A.H.; Formal analysis, H.D.; Investigation, H.D.; Resources, R.W.P.; Data curation, H.D.; Writing—original draft preparation, H.D.; Writing—review and editing, R.W.P.; Visualization, H.D.; Supervision, R.W.P.; Funding acquisition, R.W.P.

Funding: This research was funded by American Chemical Society—Petroleum Research Fund grant number PRF # 56918-ND9.

Acknowledgments: Acknowledgment is made to the donors of The American Chemical Society Petroleum Research Fund for support of this research.

Conflicts of Interest: The authors declare no conflict of interest.

References

1. Pitz, R.W.; Hu, S.T.; Wang, P.Y. Tubular premixed and diffusion flames: Effect of stretch and curvature. *Prog. Energy Combust. Sci.* **2014**, *42*, 1–34. [CrossRef]
2. Hall, C.A.; Kulatilaka, W.D.; Gord, J.R.; Pitz, R.W. Quantitative atomic hydrogen measurements in premixed hydrogen tubular flames. *Combust. Flame* **2014**, *161*, 2924–2932. [CrossRef]
3. Hall, C.A.; Pitz, R.W. Numerical simulation of premixed H_2–air cellular tubular flames. *Combust. Theory Model.* **2016**, *20*, 328–348. [CrossRef]
4. Mosbacher, D.M.; Wehrmeyer, J.A.; Pitz, R.W.; Sung, C.J.; Byrd, J.L. Experimental and numerical investigation of premixed tubular flames. *Proc. Combust. Inst.* **2002**, *29*, 1479–1486. [CrossRef]
5. Wang, Y.; Hu, S.T.; Pitz, R.W. Extinction and cellular instability of premixed tubular flames. *Proc. Combust. Inst.* **2009**, *32*, 1141–1147. [CrossRef]
6. Shopoff, S.W.; Wang, P.Y.; Pitz, R.W. The effect of stretch on cellular formation in non-premixed opposed-flow tubular flames. *Combust. Flame* **2011**, *158*, 876–884. [CrossRef]
7. Shopoff, S.W.; Wang, P.Y.; Pitz, R.W. Experimental study of cellular instability and extinction of non-premixed opposed-flow tubular flames. *Combust. Flame* **2011**, *158*, 2165–2177. [CrossRef]
8. Hall, C.A.; Pitz, R.W. Major species investigation of non-premixed cellular tubular flame. In Proceedings of the 54th AIAA Aerospace Sciences Meeting, San Diego, CA, USA, 4–8 January 2016; Number 1203.
9. Hall, C.A.; Pitz, R.W. Experimental and numerical study of H_2-air non-premixed cellular tubular flames. *Proc. Combust. Inst.* **2017**, *36*, 1595–1602. [CrossRef]
10. Wang, P.Y.; Hu, S.T.; Pitz, R.W. Numerical investigation of the curvature effects on diffusion flames. *Proc. Combust. Inst.* **2007**, *31*, 989–996. [CrossRef]
11. Hu, S.T.; Pitz, R.W.; Wang, Y. Extinction and near-extinction instability of non-premixed tubular flames. *Combust. Flame* **2009**, *156*, 90–98. [CrossRef]
12. Tinker, D.C.; Hall, C.A.; Pitz, R.W. Measurement and simulation of partially-premixed cellular tubular flames. *Proc. Combust. Inst.* **2019**, *37*, 2021–2028. [CrossRef]
13. Hu, S.T.; Pitz, R.W. Structural study of non-premixed tubular hydrocarbon flames. *Combust. Flame* **2009**, *156*, 51–61. [CrossRef]
14. Li, Q.; Fernandez, L.; Zhang, P.Y.; Wang, P.Y. Stretch and curvature effects on NO emission of H_2/air diffusion flames. *Combust. Sci. Technol.* **2015**, *187*, 1520–1541. [CrossRef]
15. Suenaga, Y.; Kitano, M.; Yanaoka, H. Extinction of cylindrical diffusion flame. *J. Thermal Sci. Technol.* **2011**, *6*, 323–332. [CrossRef]
16. Suenaga, Y.; Yanaoka, H.; Momotori, D. Influences of stretch and curvature on the temperature of stretched cylindrical diffusion flames. *J. Thermal Sci. Technol.* **2016**, *11*, JTST0028. [CrossRef]
17. Nishioka, M.; Inagaki, K.; Ishizuka, S.; Takeno, T. Effects of pressure on structure and extinction of tubular flame. *Combust. Flame* **1991**, *86*, 90–100. [CrossRef]
18. Glassman, I.; Yetter, R.A.; Glumac, N.G. *Combustion*, 5th ed.; Academic Press: Cambridge, MA, USA, 2015.
19. Correa, S.M. A review of NO_x formation under gas-turbine combustion conditions. *Combust. Sci. Technol.* **1993**, *87*, 329–362. [CrossRef]
20. Park, J.; Kim, K.T.; Park, J.S.; Kim, J.S.; Kim, S.; Kim, T.K. A study on H_2-air counterflow flames in highly preheated air diluted with CO_2. *Energy Fuels* **2005**, *19*, 2254–2260. [CrossRef]
21. Lim, J.; Gore, J.; Viskanta, R. A study of the effects of air preheat on the structure of methane/air counterflow diffusion flames. *Combust. Flame* **2000**, *121*, 262–274. [CrossRef]
22. Rørtveit, G.J.; Hustad, J.E.; Li, S.C.; Williams, F.A. Effects of diluents on NO_x formation in hydrogen counterflow flames. *Combust. Flame* **2002**, *130*, 48–61. [CrossRef]
23. Yang, K.H.; Shih, H.Y. NO formation of opposed-jet syngas diffusion flames: Strain rate and dilution effects. *Int. Hydrog. Energy* **2017**, *42*, 24517–24531. [CrossRef]
24. Shih, H.Y.; Hsu, J.R. Computed NO_x emission characteristics of opposed-jet syngas diffusion flames. *Combust. Flame* **2012**, *159*, 1851–1863. [CrossRef]
25. Hu, S.T.; Wang, P.Y.; Pitz, R.W.; Smooke, M.D. Experimental and numerical investigation of non-premixed tubular flames. *Proc. Combust. Inst.* **2007**, *31*, 1093–1099. [CrossRef]
26. Schlichting, H. *Boundary-Layer Theory*, 7th ed.; McGraw-Hill: New York, NY, USA, 1979.

27. Wang, P.Y.; Wehrmeyer, J.A.; Pitz, R.W. Stretch rate of tubular premixed flames. *Combust. Flame* **2006**, *145*, 401–414. [CrossRef]

28. Matalon, M. On flame stretch. *Combust. Sci. Technol.* **1983**, *31*, 169–181. [CrossRef]

29. Smith, G.P.; Golden, D.M.; Frenklach, M.; Moriarty, N.W.; Eiteneer, B.; Goldenberg, M.; Bowman, C.T.; Hanson, R.K.; Song, S.; Gardiner, W.C., Jr.; et al. GRI-Mech 3.0. 2000. Available online: http://www.me.berkeley.edu/gri_mech (accessed on 10 October 2018).

30. Grcar, J.F.; Day, M.S.; Bell, J.B. *Conditional and Opposed Reaction Path Diagrams for the Analysis of Fluid-Chemistry Interactions*; LBNL Report No. 52164; Lawrence Berkeley National Laboratory: Berkeley, CA, USA, 2003.

31. Warnatz, J.; Maas, U.; Dibble, R.W. *Combustion*, 4th ed.; Springer: Berlin/Heidelberg, Germany, 2006.

32. Brown, T.M.; Tanoff, M.A.; Osborne, R.J.; Pitz, R.W.; Smoke, M.D. Experimental and numerical investigation of laminar hydrogen-air counterflow diffusion flames. *Combust. Sci. Technol.* **1997**, *129*, 71–88. [CrossRef]

energies

MDPI

Article

General Correlations of Iso-octane Turbulent Burning Velocities Relevant to Spark Ignition Engines

Minh Tien Nguyen, Dewei Yu, Chunyen Chen and Shenqyang (Steven) Shy *

Department of Mechanical Engineering, National Central University, Tao-yuan City 32001, Taiwan; 103383604@cc.ncu.edu.tw (M.T.N.); 104323052@cc.ncu.edu.tw (D.Y.); 105323086@cc.ncu.edu.tw (C.C.)
* Correspondence: sshy@ncu.edu.tw

Received: 10 April 2019; Accepted: 12 May 2019; Published: 15 May 2019

Abstract: A better understanding of turbulent premixed flame propagation is the key for improving the efficiency of fuel consumption and reducing the emissions of spark ignition gasoline engines. In this study, we measure turbulent burning velocities (S_T) of pre-vaporized iso-octane/air mixtures over wide ranges of the equivalence ratio ($\phi = 0.9–1.25$, $Le \approx 2.94–0.93$), the root-mean-square (r.m.s.) turbulent fluctuating velocity ($u' = 0–4.2$ m/s), pressure $p = 1–5$ atm at $T = 358$ K and $p = 0.5–3$ atm at $T = 373$ K, where Le is the effective Lewis number. Results show that at any fixed p, T and u', $Le < 1$ flames propagate faster than $Le > 1$ flames, of which the normalized iso-octane S_T/S_L data versus u'/S_L are very scattering, where S_L is the laminar burning velocity. But when the effect of Le is properly considered in some scaling parameters used in previous correlations, these large scattering iso-octane S_T/S_L data can be collapsed onto single curves by several modified general correlations, regardless of different ϕ, Le, T, p, and u', showing self-similar propagation of turbulent spherical flames. The uncertainty analysis of these modified general correlations is also discussed.

Keywords: iso-octane; high-pressure turbulent burning velocity; Lewis number; general correlations; self-similar spherical flame propagation

---

1. Introduction

Designing a better spark ignition (SI) gasoline engine with higher thermal efficiency and lower engine emissions necessitates a thorough understanding of flame kernel initiation and its subsequent flame propagation that take place under high pressure (p), high temperature (T), high r.m.s. turbulent fluctuating velocity (u') and other complex conditions. For example, the mixture characteristics in the vicinity of an engine top dead center are frequently changed from lean to stoichiometry or even rich when the loads vary from low to high [1]. Besides, the mixture characteristics and values of u' change from cylinder to cylinder, resulting in the difference of flame initiation and subsequent flame propagation speed. Therefore, the turbulent burning velocity (S_T) has been introduced for further understanding of turbulent premixed flame propagation [2]. However, S_T data for liquid fuels (e.g., iso-octane, the major component of gasoline surrogate) under SI engine relevant conditions are rare, since most studies applied gaseous fuels (e.g., methane, propane, hydrogen at atmospheric condition [2–7] or at elevated pressure and room temperature conditions [8–16]). Whether these gaseous S_T data could be applicable to SI gasoline engines is still an open issue. To the best knowledge of the authors, the only available iso-octane S_T data which satisfied high-T, -p and -u' conditions were that reported by Lawes et al. [17] and Nguyen et al. [18] using turbulent spherical expanding flame at $T = 360$ K [17] and 423 K [18]. Clearly, more lean and rich iso-octane S_T data under high-T, -p, and -u' conditions are needed in order to make a detailed comparison with previous gaseous S_T data and thus address the question of similarity and difference between liquid iso-octane and gaseous data, as the first objective of this study.

Lean and rich iso-octane/air mixtures have very different Lewis number (*Le*). For instance, the lean iso-octane/air mixture at the equivalence ratio $\phi = 0.9$ has $Le \approx 2.94 >> 1$, while the rich iso-octane/air mixture at $\phi = 1.25$ has $Le \approx 0.93 < 1$. Note that *Le* is the effective Lewis number which is estimated by the ratio between thermal diffusivity and mass diffusivity with the mass diffusivity being that of the deficient reactants and the abundant inert, i.e., $Le \approx Le_{fuel}$ for lean mixture and $Le \approx Le_{oxygen}$ for the rich mixture [17]. Hence, the second objective of this paper is to seek possible general correlations of S_T with the consideration of the *Le* effect to represent the liquid iso-octane/air mixture at lean and rich conditions in attempt to enhance our understanding of high-pressure/temperature turbulent flame propagation that may be relevant to SI gasoline engines.

Seeking a general correlation of iso-octane S_T data is of high interest for SI engines using liquid fuels, and it is still an unresolved problem. Due to the complexity of turbulence-chemistry interactions, the different configuration of experimental facilities, and the uncertainty of measurement techniques, a general correlation to predict S_T is not always clear despite many S_T correlations available in literatures (e.g., [2–4,8–10,12–24], among others). Thus, in this work we make our efforts to analyze the scaling parameters in some selected correlations by taking the effect of *Le* into consideration based on the present measured S_T data using a single liquid fuel (iso-octane). Therefore, proper scaling parameters including the effect of *Le* and better modified general correlations using the same liquid iso-octane S_T data may be obtained. In this study, five general correlations, all including the effect of *Le*, are considered and tested by using the current measured iso-octane S_T data. Further, a mean absolute percentage error (MAPE) [23] is used to assess the accuracy of these five correlations. The first correlation is the Bradley's correlation [3]:

$$S_{T,\bar{c} = 0.5}/u' \sim (KLe)^{-0.3} \tag{1}$$

$\bar{c}$ is the mean progress variable, i.e., $\bar{c} = 0$ and $\bar{c} = 1$ represent reactant and product, respectively. The stretch factor $K \approx 0.25(u'/S_L)^2(Re_T)^{-0.5}$ [25], where the turbulent flow Reynolds number $Re_T = u'L_I/\nu$, L_I is the integral length scale, and ν is the kinematic viscosity of reactant. Second, by grouping the scaling parameters in correlations proposed by Kobayashi et al. [19], Chaudhuri et al. [12], and Shy et al. [16] with Le^{-1}, three modified correlations have been reported in Reference [18], respectively as shown below:

$$S_{T,\bar{c} = 0.5}/S_L \sim [(u'/S_L)(p/p_0)Le^{-1}]^{0.42}, \text{ where } p_0 = 1 \text{ atm} \tag{2}$$

$$S_{T,\bar{c} = 0.5}/S_L \sim (Re_{T,flame}Le^{-1})^{0.5}, \text{ where } Re_{T,flame} = (u'\langle R \rangle)/\alpha = (u'\langle R \rangle)/(\delta_L S_L) \tag{3}$$

$$S_{T,\bar{c} = 0.5}/u' \sim (DaLe^{-1})^{0.5}, \text{ where } Da = (L_I/u')(S_L/\delta_L) \tag{4}$$

$\langle R \rangle$ is the mean turbulent spherical flame radius, δ_L is the laminar flame thickness, and $\alpha = \delta_L S_L$ is the thermal diffusivity. In Equation (3), the commonly-used integral length scale of turbulence is replaced by the average flame radius $\langle R \rangle$ and the kinematic viscosity (ν) is replaced by the thermal diffusivity ($\alpha \approx \delta_L S_L$), as proposed by Chaudhuri et al. [12,13]. Finally, the fifth correlation is that reported by Ritzinger [20]:

$$S_{T,\bar{c} = 0.5}/S_L = 1 + (0.46/Le^{0.85})(u'/S_L)^{0.5Le^{0.25}}(Re_T)^{0.25}(p/0.1\text{MPa})^{0.1} \tag{5}$$

Equation (5) has been tested against methane/air S_T data at high-pressure gas engine conditions [26], while Equations (1)–(4) have not yet done so. To see how well these five correlations, Equations (1)–(5), could be applicable to SI engines for S_T prediction, an uncertainty comparison is performed by accessing the accuracy of these correlations based on the MAPE value which is calculated using the following equation [23].

$$\text{MAPE}(\%) = \frac{\sum_{i=1}^{n} \left| \frac{x_{exp,i} - x_{corr,i}}{x_{exp,i}} \right|}{n} \times 100 \tag{6}$$

$x_{exp,i}$ is the experimental value, $x_{corr,i}$ is the value predicted by the correlation, and n is the number of the data group under examination.

In short, this work measures S_T data of lean, stoichiometric, and rich iso-octane/air mixtures with $Le > 1$ and $Le < 1$ using a large dual-chamber, constant pressure/temperature, fan-stirred cruciform explosion facility, capable of generating near-isotropic turbulence. It will be shown in due course that all the scattering S_T data of iso-octane/air mixtures can be collapsed onto single curves using Equations (1)–(5). Further, the current iso-octane data are also compared with previous data proposed by Lawes et al. [17]. Finally, the uncertainty of these correlations based on the MAPE is discussed.

2. Experimental Method

Experiments are conducted in a large dual-chamber, constant-temperature, constant-pressure, fan-stirred cruciform explosion facility for S_L and S_T measurements of expanding spherical iso-octane/air flames. The reader is directed to [18,21] and references therein for detail treatment on the facility and its associated turbulence properties. For completeness, a simplified sketch of the 3D cruciform burner resided in a large pressure vessel with optical accesses is added, as shown in Figure 1, alongside the Schlieren imaging arrangement. Below are the descriptions of the experimental procedures.

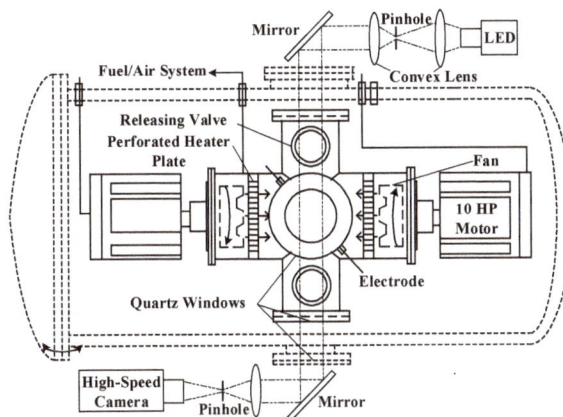

Figure 1. The high-pressure/temperature, double-chamber explosion facility for premixed turbulent combustion studies [27].

Before a run, we first vacuum the heated 3D cruciform burner (Figure 1) and then inject the appropriate mole fractions of pre-vaporized iso-octane and air into the burner to the desired initial pressure and temperature conditions. Two counter-rotating fans at a frequency of 30 Hz are turned on to well mix the iso-octane/air mixture for 4 minutes and reach a uniform temperature distribution with less than 1 °C variation in the domain of experimentation. The temperature uniformity is achieved by using effective turbulent heat convection through a pair of heated perforated plates (please see [18,21] for details). The domain of experimentation is set at $0.17 \leq \langle R \rangle / R_{min} \leq 0.30$ to avoid the ignition influence at the early stage of kernel development and the wall effects at the later stage of flame propagation, where the minimum wall confinement radius of the 3D cruciform bomb R_{min} is about 150 mm. For laminar flame speed measurements, we ignite the mixtures shortly (about 10 s) after the fan turn-off to allow for the decay of turbulence to quiescence, while keeping a uniform temperature distribution in the experimental domain [21]. Centrally-ignited flame initiation and its subsequent flame propagation are recorded by the high-speed Schlieren imaging to obtain the time evolution of the average flame radii $\langle R \rangle(t) = [A(t)/\pi]^{0.5}$, where $A(t)$ is the area enclosed by the laminar/turbulent flame front. The flame speed $d\langle R \rangle/dt$ is obtained by the central differentiation of $\langle R \rangle$ vs. t data.

A typical refined gasoline consists of hundreds of different components that may vary batch by batch depending on the source of crude oil and the refinery processes, making the benchmark of S_T for gasoline difficult to measure accurately. For simplicity, iso-octane, a major surrogate component for gasoline, is used to gain insight and understanding of the underlying physical mechanism of turbulent premixed flame propagation. As such, this work measures values of S_T for iso-octane/air mixtures as a function of ϕ, u', p, T, and Le. Table 1 lists laminar properties of the iso-octane/air mixture at various conditions including T, ϕ, Le, p and corresponding laminar burning velocities (S_L). It should be noted that the selected mixtures, i.e., $\phi = 0.9$ & 1.25 at $T = 358$K at $p = 1$ atm and $\phi = 1.0$ & 1.2 at $T = 373$ K at $p = 1$ atm, have closely matched values of S_L (see Table 1) but with different Le ($Le > 1$ and $Le < 1$). As such, the effect of Le on S_T of iso-octane/air mixtures can be studied.

Table 1. Conditions and properties of iso-octane/air mixtures.

T (K)	ϕ	Le	p (atm)	S_L (cm/s)
358	0.9/1.0/1.25	2.94/1.43/0.93	1, 3, 5	41.0, 26.2, 24.8/45.0, 31.2, 27.6/40.2, 32.9, 28.2
373	1.0/1.2	1.43/ 0.93	0.5, 1, 3	49.0, 44.9, 31.0/47.0, 44.6, 36.0
423 [1]	1.0	1.43	1, 3, 5	55.0, 42.0, 36.0

[1] Previous data at 423 K [18].

3. High-Speed Schlieren Imaging of Turbulent Spherical Flame and Their S_T Determination

Figure 2 shows the effects of Le and p on the emergence of small scale structures and the increase of average flame propagation rate at fixed u' with increasing p for both (a) laminar and (b) turbulent cases, where all images have the same $\langle R \rangle \approx 35$ mm. At any fixed p, the elapsed instants after ignition show that $Le < 1$ flames (2nd and 4th columns) propagate much faster than $Le > 1$ flames (1st and 3rd columns). S_L decreases with increasing p even though the flame at 5 atm appears more wrinkling due to the emergence of cellular structures allover the flame surface especially for the case at 5 atm and $Le \approx 0.93 < 1$ (see the 4th column of the first row images in Figure 2a). For the turbulent case at constant $u' = 1.4$ m/s (Figure 2b), the turbulent flame propagates faster with increasing p under both $Le > 1$ and $Le < 1$ conditions, where the turbulent wrinkled flame at 5 atm are full of very small scale structures. These fine structures are mainly due to the reduction of the thickness of the laminar flamelet at high pressure that promotes the hydrodynamic instability and thus increases the wrinkled flame front surface, but they contribute only a small part for the increase of S_T. This is because when the flow turbulent Reynolds number ($Re_{T,flow} = u'L_I/\nu$) is kept constant for gaseous fuels (e.g., methane and syngas), S_T actually decreases with increasing pressure, similar to S_L, showing a global response of burning velocities to the increase of pressure (please see Reference [11] for detail information). As such, the increase of S_T with increasing p at fixed u' is probably mainly due to the increase of $Re_{T,flow}$ at elevated pressure because $\nu \sim \rho^{-1} \sim p^{-1}$.

Figure 3a shows the averaged flame radii of iso-octane/air turbulent premixed flames versus time for $Le > 1$ ($\phi = 0.9$ and 1.0) and $Le < 1$ ($\phi = 1.25$) at the same $p = 3$ atm, $u' = 1.4$ m/s, and $T = 358$ K conditions. Flame speeds, $d\langle R \rangle(t)/dt$ and S_F or S_T, can be estimated from the raw data of $\langle R \rangle(t)$ within the range of 25 mm $\leq \langle R \rangle \leq 45$ mm. $d\langle R \rangle/dt$ is directly taking the time differentiation on $\langle R \rangle(t)$, while S_F is determined as the slope of the best linear-fit of $\langle R \rangle(t)$ within 25 mm $\leq \langle R \rangle \leq 45$ mm. Within this experimentation domain, S_F is just the average value of the linear increase $d\langle R \rangle/dt$ data ($S_F = \overline{dR/dt}$), as substantiated in Figure 3b. Note that these iso-octane data of $\langle R \rangle$ vs. t and $d\langle R \rangle/dt$ vs. t or $\langle R \rangle$ in Figure 3 are very similar to previous data obtained from the Leeds fan-stirred explosion bomb [17,28]. It has been shown in Reference [11] that the modified S_F using the density correction and Bradley's mean progress variable $\bar{c}$ converting factor for Schlieren spherical flames at $\bar{c} = 0.5$ [25], i.e., $S_{T,\bar{c}=0.5} \approx (\rho_b/\rho_u)S_F(\langle R \rangle_{\bar{c}=0.1}/\langle R \rangle_{\bar{c}=0.5})^2 = (\rho_b/\rho_u)\overline{dR/dt}(\langle R \rangle_{\bar{c}=0.1}/\langle R \rangle_{\bar{c}=0.5})^2$, show good agreements between Bunsen-type and spherical flames, suggesting that S_T determined at the flame surface contour of $\bar{c} = 0.5$ may be a better representative of itself regardless of different flame geometries. The subscripts b and u indicate burned and unburned gases. For Schlieren expanding spherical turbulent premixed flames, $\langle R \rangle_{\bar{c}=0.1}/\langle R \rangle_{\bar{c}=0.5} \approx 1.4$ [22,28]. In this work, all measured S_F data are converted to turbulent burning velocities at $\bar{c} = 0.5$ ($S_{T,\bar{c}=0.5}$).

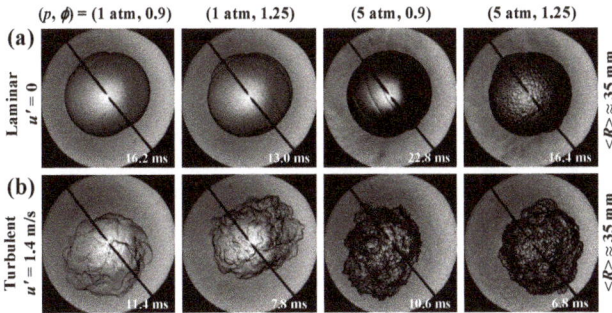

Figure 2. High-speed Schlieren images of lean and rich iso-octane/air expanding spherical premixed flames, i.e., $\phi = 0.9$ with $Le \approx 2.94 > 1$ and $\phi = 1.25$ with $Le \approx 0.93 < 1$ at 358 K, showing effects of Lewis number and pressure on the emergence of small scale structures and the increase in average flame propagation rate at fixed $u' = 1.4$ m/s where the flow turbulent Reynolds number ($Re_{T,flow} = u'L_I/\nu$) increases with pressure because $\nu \sim \rho^{-1} \sim p^{-1}$. All images in both (a) laminar and (b) turbulent cases have the same $\langle R \rangle \approx 35$mm in the same view field of 110 mm × 110 mm for comparison.

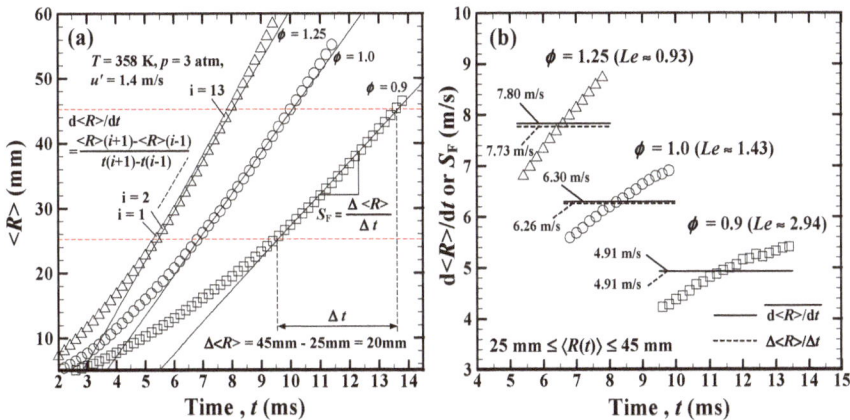

Figure 3. (a) Typical averaged flame radii of turbulent iso-octane/air premixed flames as a function of time at three different values of ϕ with $Le > 1$ ($\phi = 0.9$ and 1.0) and $Le < 1$ ($\phi = 1.25$) under the same $p = 3$ atm, $u' = 1.4$ m/s and $T = 358$ K conditions. (b) Two turbulent flame speeds versus time, $d\langle R \rangle/dt$ and S_F (the slope of $\langle R \rangle(t)$) as indicated in (a), which are almost equal to each other within 25 mm $\leq \langle R(t) \rangle \leq$ 45 mm.

4. Results and Discussion

Figure 4a presents laminar and turbulent burning velocities of iso-octane/air mixtures as a function of p for three different values of ϕ with different Le varying from 0.93 to 2.94. As increase p, values of S_L decrease in a minus power law manner of the form $S_L \sim p^{-n_L}$, where the exponent n_L increases with increasing Le from $n_L = 0.14$ at $Le \approx 0.93$ and $n_L = 0.31$ at $Le \approx 1.43$ to $n_L = 0.33$ at $Le \approx 2.94$. This suggests that $Le < 1$ flames are less sensitive to pressure elevation than $Le > 1$ flames. On the other hand, values of $S_{T,\bar{c}=0.5}$ increase with increasing p in a positive power law form, i.e., $S_T \sim p^{+n_T}$, where n_T decreases with increasing Le from $n_T = 0.14$ at $Le \approx 0.93$ and $n_T = 0.1$ at $Le \approx 1.43$ to $n_T = 0.03$ at $Le \approx 2.94$. The latter case for the lean iso-octane/air mixture at $\phi = 0.9$ shows that $S_T \sim p^{0.03}$ when $u' = 1.4$ m/s. In other words, at large $Le \approx 2.94 >> 1$ and $u' = 1.4$ m/s, the fine structures allover the flame front surface (Figure 2b) induced by the hydrodynamic instability at high pressure as well as the increase of $Re_{T,flow}$ due to the decrease of kinematic viscosity at elevated pressure only have a rather small influence on S_T. The reason for this tiny impact on S_T due to the increase of p when $u' = 1.4$ m/s is not yet clear, which deserves further studies. Figure 4b presents laminar and turbulent burning velocities as a function of temperature at four different values of u' for the stoichiometric iso-octane/air mixture at 1 atm. In general, values of S_L and $S_{T,\bar{c} = 0.5}$ increase with increasing T. $S_{T,\bar{c} = 0.5} \sim (T/T_0)^{m_T}$, where $T_0 = 298$ K. The exponent constant $m_T = 0.07$ is small at small $u' = 1.4$ m/s, while the value of m_T increases to 1.07 when $u' = 4.2$ m/s. This suggests that $S_{T,\bar{c} = 0.5}$ is not-so-sensitive to the increase of temperature at $u' = 1.4$ m/s in line with previous studies by Lipatnikov et al. [29] and Fogla et al. [30] who studied the effect of density ratio on the turbulent flame speed. Note that changing the temperature is equivalent to changing the density ratio between the burned and unburned gases that has insignificant effect on S_T at low-to-moderate turbulent intensities [29,30]. However, such temperature enhancement sensitivity increases with increasing u'.

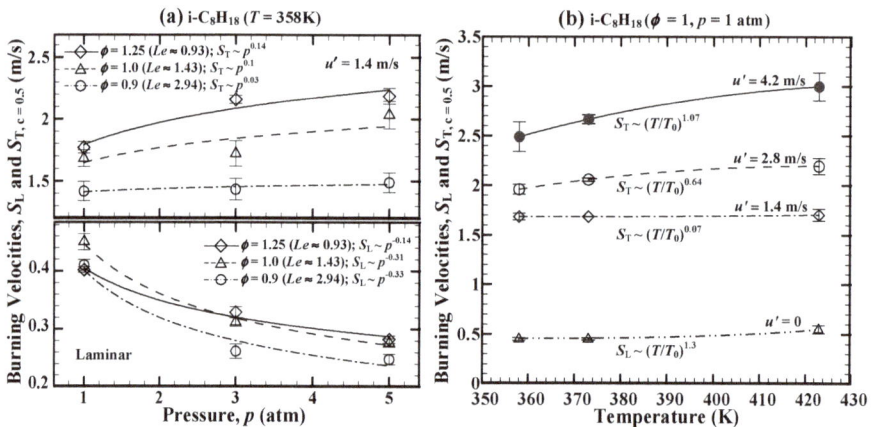

Figure 4. (**a**) Laminar burning velocities and turbulent burning velocities at $\bar{c} = 0.5$ of the iso-octane/air mixtures at three different ϕ with $Le < 1$ and $Le > 1$ as a function of pressure, where $T = 358$ K and $u' = 1.4$ m/s. (**b**) Effect of temperature on S_L and S_T at $\bar{c} = 0.5$ having four different u' varying from 0 to 4.2 m/s.

The current S_T data obtained at 373 K and 5 atm may not be sufficiently high, since much higher temperatures and pressures are typically observed in SI engines. However, the available S_T data of pre-vaporized iso-octane fuel (the major component of gasoline surrogate) even at 373 K and 5 atm are still very rare, which may be important as the first order approximation towards the practical SI engine conditions. Moreover, these experimental results are useful because they can be used to improve current models for the predication of turbulent flame speeds of pre-vaporized iso-octane fuel at high p and high T.

All present $S_{T,\bar{c}=0.5}/S_L$ of iso-octane/air mixtures at $\phi = 0.9$–1.25 at $T = 358$ K and/or 373 K together with previous data at $\phi = 1.0$ and at $T = 423$ K [18] are grouped onto two data sets having $Le < 1$ and $Le > 1$ and plotted against turbulent intensities (u'/S_L), as shown in Figure 5. Results show that $Le < 1$ flames with $S_{T,\bar{c}=0.5}/S_L = 2.54(u'/S_L)^{0.53}$ having a poor goodness $R^2 = 0.58$ propagates faster than $Le > 1$ flames with $S_{T,\bar{c}=0.5}/S_L = 1.76(u'/S_L)^{0.64}$ and $R^2 = 0.65$, revealing large data scattering. Note that the increase of $S_{T,\bar{c}=0.5}/S_L$ with u'/S_L is not linear. There is a bending effect of $S_{T,\bar{c}=0.5}/S_L$ at higher u'/S_L for both $Le < 1$ and $Le > 1$ flames. Such bending effect has been discussed by many studies (e.g., [2,4,11,19,22,25] among others), which is mainly attributed to the turbulent flame stretch effect.

At higher u'/S_L, the two fitting curves are getting closer to each other, suggesting that the effect of Le is getting weaker in intense turbulence. Nevertheless, it is obvious that the Le effect should be considered in any scaling correlations.

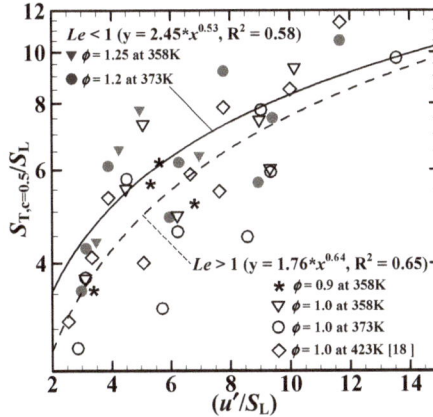

Figure 5. Normalized turbulent burning velocities at $\bar{c} = 0.5$ plotted against turbulent intensities (u'/SL) at different pressure $p = 0.5$–3 atm at 373 K and $p = 1$–5 atm at 358 K and/or 423 K.

First, we discuss the correlation of Equation (1) proposed by Bradley et al. [3] which is used to fit all data from Figure 5. The result is presented in Figure 6, suggesting that the Bradley's correlation can be used to fit all scattering S_T data for $Le > 1$ and $Le < 1$ flames with a better goodness of $R^2 = 0.77$ than that in Figure 5. In [3], Bradley et al. noted that Equation (1) was limited within the range of KLe between 0.01 and 0.63 and, if outside that range, the logarithmic plots of u_t/u'_k (or S_T/u') against KLe has large data scattering (please see Reference [3]). Also plotted in Figure 6 for comparison is the original version of Equation (1): $S_{T,\bar{c}=0.5}/u' = 0.88\,(KLe)^{-0.3}$ based on various gaseous mixtures [3]. The present pre-factor 0.55 is 62.5% of the previous pre-factor of 0.88 as in [3], probably due to different liquid and gaseous fuels applied.

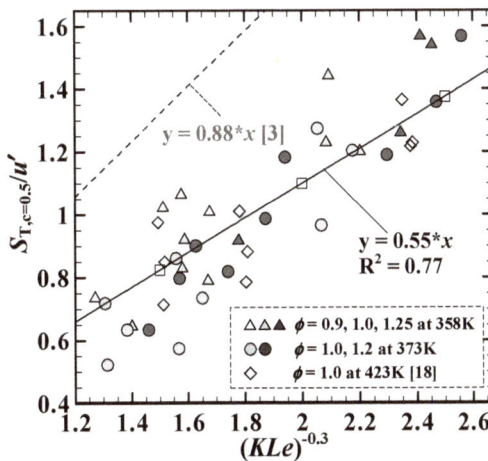

Figure 6. Normalized turbulent burning velocities of iso-octane/air mixtures using the Bradley's correlation [3]: $S_{T,\bar{c}=0.5}/u' = 0.55(KLe)^{-0.3}$. The dashed line was the original version of previous data obtained by Bradley et al. [3], where $S_{T,\bar{c}=0.5}/u' = 0.88(KLe)^{-0.3}$.

Figure 7a–c show the normalized $S_{T,\bar{c}=0.5}$ data by their corresponding values of S_L and/or u' for iso-octane/air mixtures with $Le > 1$ and $Le < 1$ using the earlier versions of Equations (2)–(4) without the consideration of Le. These three earlier versions of scaling parameters are respectively: (a) $(u'/S_L)(p/p_0)$ proposed by Kobayashi et al. [8,19], (b) $(Re_{T,flame})^{0.5}$ by Chaudhuri et al. [12], and (c) $(Da)^{0.5}$ by Shy et al. [10,11,16]. The results clearly show two distinct data sets with $Le < 1$ flames (dark grey symbols) propagating faster than $Le > 1$ flames (white and light grey symbols) for all three correlations as shown in Figure 7a–c. Clearly, the effect of Le must be taken into the consideration to obtain possible general correlations. When the aforesaid three scaling parameters are grouped with Le^{-1}, all present iso-octane $S_{T,\bar{c}=0.5}$ data can be collapsed onto single curves, as shown in Figure 7d–f, suggesting that the propagation of iso-octane turbulent premixed spherical flames is self-similar, regardless of different Le, ϕ, p, and u' applied. The goodness between these three modified correlations varies from $R^2 \approx 0.89$ using Equation (2) and $R^2 \approx 0.84$ using Equation (3) to $R^2 \approx 0.7$ using Equation (3), all based on the same iso-octane/air $S_{T,\bar{c}=0.5}$ data.

The final $S_{T,\bar{c}=0.5}/S_L$ correlation to be discussed is Equation (5) proposed by Ritzinger [20] based on a complicated empirical relation in the form of $S_{T,\bar{c}=0.5}/S_L = 1 + (0.46/Le^{0.85})(u'/S_L)^{0.5Le^{0.25}}(Re_T)^{0.25}(p/0.1\text{MPa})^{0.1}$. Results are presented in Figure 8, where all the present iso-octane $S_{T,\bar{c}=0.5}$ data can be well collapsed onto a single curve with $R^2 = 0.89$. In general, Equations (1)–(5) can be used as general correlations for both present liquid iso-octane/air and preveous gaseous turbulent burning velocities with reasonably good R^2, regardless of different fuel, ϕ, Le, u', T and p applied.

Here we apply Equation (6) to estimate MAPE and the results are shown in Figure 9. MAPE is a better statistical variable than the goodness R^2, because the latter depends strongly on the available data volume to estimate the uncertainty of these five general correlations (Equations (1)–(5)). Note that the lower the MAPE is, the more accuracy of the correlation is. All MAPE percentages for the aforesaid five general correlations are less than 13%, showing that these general correlations are reasonably good. Among them, the best is Equation (5) proposed by Ritzinger [20] with a lowest MAPE of 8.5% and the second best is Equation (2) proposed by Kobayashi et al. [8,19] with a MAPE of 9.6%. The other three general correlations (Equations (1), (3), (4)) have roughly the same MAPE (11.9%–12.9%). It should be noted that Equation (2) is probably the most convenient one to use, because it does not require any length scales of turbulence and flame.

We now discuss the iso-octane comparison between the present data and previous data obtained by Lawes et al. [17] using Equation (4). The comparison results are presented in Figure 10. The main reason that we use Equation (4) for comparison is because the Damköhler number (Da) to the one half power has been extensively used by Peters [31]. Besides, $Da = (L_I/u')(S_L/\delta_L)$ includes the important integral length scale of turbulence (L_I), but Equation (2) and Equation (3) do not include L_I. A good agreement between present and previous data is found. Both present and previous iso-octane $S_{T,\bar{c}=0.5}$ data can be represented by a scaling relation of $S_{T,\bar{c}=0.5}/u' = 0.082\,(DaLe^{-1})^{0.5}$ with a reasonable goodness of $R^2 = 0.78$.

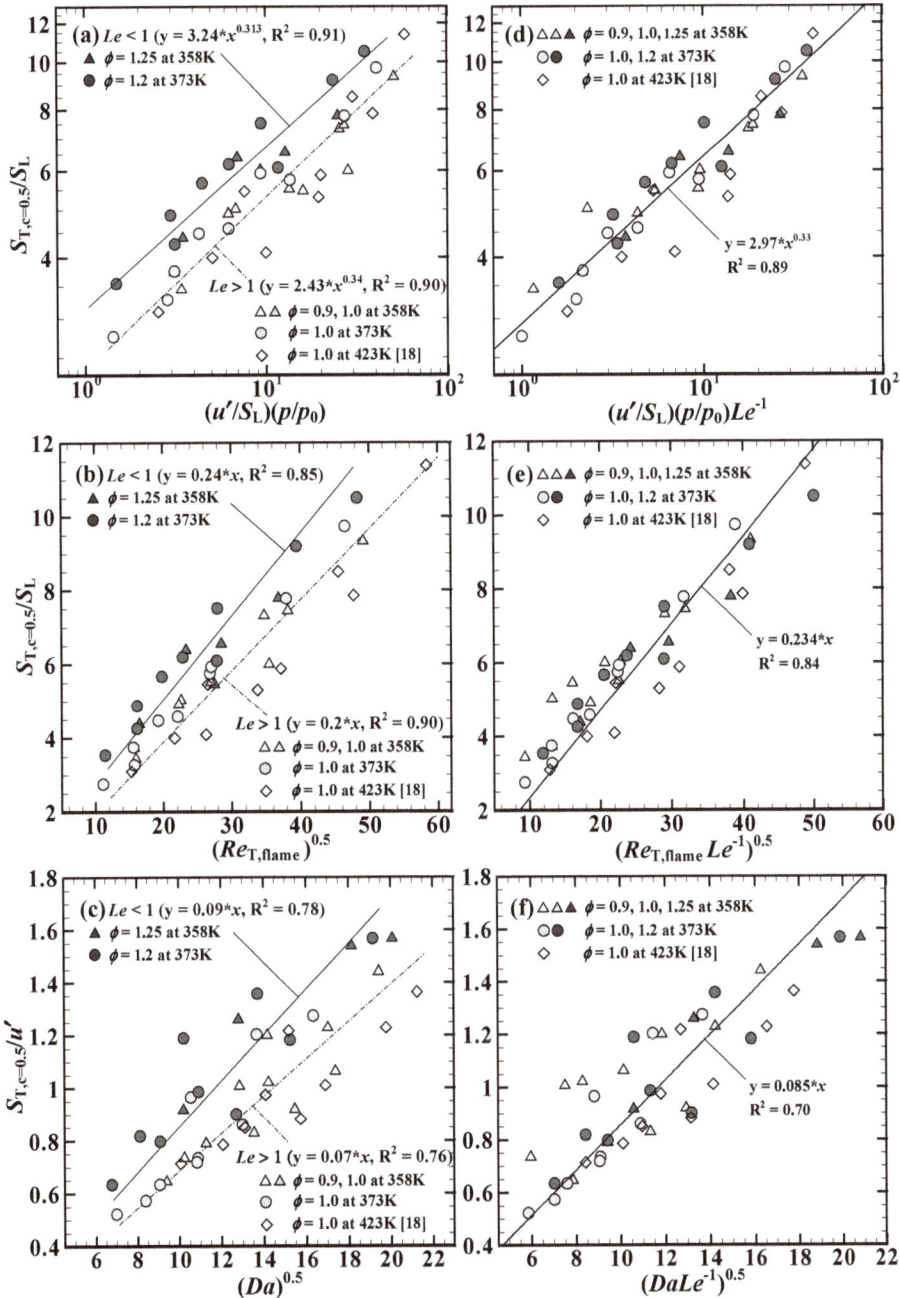

Figure 7. Left column: Normalized turbulent burning velocities for iso-octane/air mixtures plotted against three different scaling parameters: (**a**) $(u'/S_L)(p/p_0)$ [8,19]; (**b**) $(Re_{T,flame})^{0.5}$ [12]; (**c**) $(Da)^{0.5}$ [10, 11,16], all showing that $Le < 1$ flames propagate faster than $Le > 1$ flames having considerable data scattering. Right column: Same data from (**a**)–(**c**), but plotted against three modified scaling parameters grouped with Le^{-1}: (**d**) $(u'/S_L)(p/p_0)Le^{-1}$ (Equation (2)); (**e**) $(Re_{T,flame}Le^{-1})^{0.5}$ (Equation (3)); (**f**) $(DaLe^{-1})^{0.5}$ (Equation (4)), in which both $Le < 1$ and $Le > 1$ data sets are collapsed onto single curves.

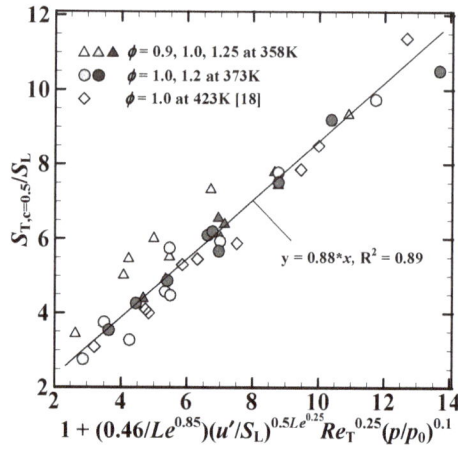

Figure 8. Normalized turbulent burning velocities of iso-octane/air mixtures using Equation (5).

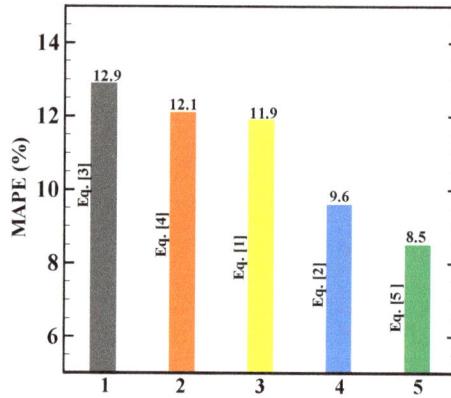

Figure 9. The accuracy of five general correlations determined through the use of mean absolute.

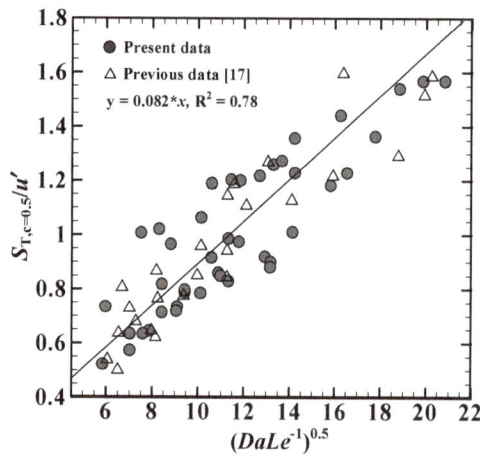

Figure 10. Comparison of present data with previous data obtained by Lawes et al. [17] using Equation (4) for the same iso-octane/air mixtures.

5. Conclusions

Turbulent burning velocities of lean, stoichiometric and rich iso-octane/air mixtures having $Le < 1$ and $Le > 1$ are measured under elevated pressures and temperatures relevant to SI gasoline engine or micro gas turbine conditions. These measurements reveal the following points:

(1) Turbulent burning velocities increase with increasing pressure and temperature at any given r.m.s turbulent fluctuating velocities. However, the increase of turbulent burning velocities is quite small at modest $u' = 1.4$ m/s, which is much less sensitive with the increase of pressure and temperature as compared to that at higher values of u'.

(2) The bending effect of $S_{T,\bar{c}=0.5}/S_L$ vs. u'/S_L curves is observed at higher u'/S_L.

(3) $Le < 1$ flames (the rich iso-octane/air mixture) propagate faster than $Le > 1$ flames (lean and/or stoichiometric iso-octane/air mixtures) at any given conditions.

(4) All present iso-octane $S_{T,c = 0.5}$ data with $Le < 1$ and $Le > 1$ can be collapsed onto single curves regardless of different T, p, u', ϕ, and Le used, which can be represented by five general correlations (Equations (1)–(5)), suggesting that turbulent premixed spherical flames have a self-similar propagation nature. These five general correlations have reasonable good accuracy due to their low values of MAPE (MAPE < 13%).

The fundamental understanding of flame initiation and propagation of lean iso-octane/air mixtures through the interactions of centrally-ignited outwardly-propagating premixed flame and intense turbulence at high pressure and high temperature is of importance for the future development of high-thermal-efficiency SI gasoline engines operated at fuel lean conditions. This is because premixed lean-burn technology can increase the mixture specific heat ratio and decrease the net heat loss to the engine cylinder through low temperature combustion, resulting in higher thermal efficiency and lower NO_x emissions [32]. As pointed out by Nakata et al. [32], super lean-burn combustion with a very strong tumble flow (very high u') in a long stroke cylinder is crucial to further enhance the maximum engine thermal efficiency up to 50% [33]. Hence, for the future research direction, the S_T information of iso-octane (the major component of gasoline surrogate) at extreme conditions, e.g., leaner mixtures ($\phi = 0.6$–0.8), higher $u' > 4.2$ m/s and higher $p > 5$ atm, are needed.

Author Contributions: Conceptualization, S.S.S.; methodology, S.S.S.; validation, M.T.N., D.Y., and C.C.; formal analysis, M.T.N., D.Y., C.C., and S.S.S.; investigation, M.T.N., D.Y., C.C., and S.S.S.; resources, S.S.S.; data curation, M.T.N., D.Y., C.C., S.S.S.; Writing—Original Draft preparation, M.T.N., S.S.S.; Writing—Review and Editing, S.S.S.; visualization, M.T.N., D.Y., and C.C.; supervision, S.S.S.; project administration, S.S.S.

Funding: This research was funded by the MINISTRY OF SCIENCE AND TECHNOLOGY, TAIWAN, grant number MOST 106-2221-E-008-054-MY3; 106-2923-E-008-004-MY3, for which we are greatly appreciated.

Conflicts of Interest: The authors declare no conflict of interest.

References

1. Tropina, A.A.; Shneider, M.N.; Miles, R.B. Ignition by short duration, nonequilibrium plasma: Basic concepts and applications in internal combustion engines. *Combust. Sci. Technol.* **2016**, *188*, 831–852. [CrossRef]

2. Abdel-Gayed, R.G.; Bradley, D.; Lawes, M. Turbulent burning velocities: A general correlation in terms of training rates. *Proc. R. Soc. A* **1987**, *414*, 389–413. [CrossRef]

3. Bradley, D.; Lau, A.K.C.; Lawes, M. Flame stretch rate as a determinant of turbulent burning velocity. *Philos. Trans. R. Soc. Lond. A* **1992**, *338*, 359–387.

4. Ronney, P.D. *Modeling in Combustion Science*; Springer: Berlin, Germany, 1994; pp. 3–20.

5. Lipatnikov, A.N.; Chomiak, J. Turbulent flame speed and thickness: Phenomenology, evaluation, and application in multi-dimensional simulations. *Prog. Energy Combust. Sci.* **2002**, *28*, 1–74. [CrossRef]

6. Lipatnikov, A.; Chomiak, J. Molecular transport effects on turbulent flame propagation and structure. *Prog. Energy Combust. Sci.* **2005**, *31*, 1–73. [CrossRef]

7. Driscoll, J.F. Turbulent premixed combustion: Flamelet structure and its effect on turbulent burning velocities. *Prog. Energy Combust. Sci.* **2008**, *34*, 91–134. [CrossRef]

8. Kobayashi, H.; Kawazoe, H. Flame instability effects on the smallest wrinkling scale and burning velocity of high-pressure turbulent premixed flames. *Proc. Combust. Inst.* **2000**, *28*, 375–382. [CrossRef]
9. Cheng, R.K.; Littlejohn, D.; Strakey, P.A.; Sidwell, T. Laboratory investigations of a low-swirl injector with H_2 and CH_4 at gas turbine conditions. *Proc. Combust. Inst.* **2009**, *32*, 3001–3009. [CrossRef]
10. Liu, C.C.; Shy, S.S.; Chen, H.C.; Peng, M.W. On interaction of centrally ignited, outwardly propagating premixed flames with fully developed isotropic turbulence at elevated pressure. *Proc. Combust. Inst.* **2011**, *33*, 1293–1299. [CrossRef]
11. Liu, C.; Shy, S.S.; Peng, M.; Chiu, C.; Dong, Y. High-pressure burning velocities measurements for centrally ignited premixed methane/air flames interacting with intense near-isotropic turbulence at constant Reynolds numbers. *Combust. Flame* **2012**, *159*, 2608–2619. [CrossRef]
12. Chaudhuri, S.; Wu, F.; Zhu, D.; Law, C.K. Flame speed and self-similar propagation of expanding turbulent premixed flames. *Phys. Rev. Lett.* **2012**, *108*, 044503. [CrossRef] [PubMed]
13. Chaudhuri, S.; Wu, F.; Law, C.K. Scaling of turbulent flame speed for expanding flames with Markstein diffusion considerations. *Phys. Rev. E* **2013**, *88*, 033005. [CrossRef]
14. Wu, F.; Saha, A.; Chaudhuri, S.; Law, C.K. Propagation speeds of expanding turbulent flames of C_4 to C_8 n-alkanes at elevated pressures: Experimental determination, fuel similarity, and stretch-affected local extinction. *Proc. Combust. Inst.* **2015**, *35*, 1501–1508. [CrossRef]
15. Venkateswaran, P.; Marshall, A.; Seitzman, J.; Lieuwen, T. Scaling turbulent flame speeds of negative Markstein length fuel blends using leading points concepts. *Combust. Flame* **2015**, *162*, 375–387. [CrossRef]
16. Shy, S.S.; Liu, C.C.; Lin, J.Y.; Chen, L.L.; Lipatnikov, A.N.; Yang, S.I. Correlations of high-pressure lean methane and syngas turbulent burning velocities: Effects of turbulent Reynolds, Damköhler, and Karlovitz numbers. *Proc. Combust. Inst.* **2015**, *35*, 1509–1516. [CrossRef]
17. Lawes, M.; Ormsby, M.P.; Sheppard, C.G.W.; Woolley, R. The turbulent burning velocity of iso-octane/air mixtures. *Combust. Flame* **2012**, *159*, 1949–1959. [CrossRef]
18. Nguyen, M.T.; Yu, D.W.; Shy, S.S. General correlations of high pressure turbulent burning velocities with the consideration of Lewis number effect. *Proc. Combust. Inst.* **2019**, *37*, 2391–2398. [CrossRef]
19. Kobayashi, H.; Seyama, K.; Hagiwara, H.; Ogami, Y. Burning velocity correlation of methane/air turbulent premixed flames at high pressure and high temperature. *Proc. Combust. Inst.* **2005**, *30*, 827–834. [CrossRef]
20. Ritzinger, J. Einfluss der Kraftstoffe RON95, Methan und Ethanol auf Flammenausbreitung und Klopfverhalten in Ottomotoren mit Abgasrückführung. Ph.D. Thesis, ETH Zürich, Zürich, Switzerland, 2013.
21. Jiang, L.J.; Shy, S.S.; Li, W.Y.; Huang, H.M.; Nguyen, M.T. High-temperature, high-pressure burning velocities of expanding turbulent premixed flames and their comparison with Bunsen-type flames. *Combust. Flame* **2016**, *172*, 173–182. [CrossRef]
22. Bradley, D.; Lawes, M.; Liu, K.; Mansour, M.S. Measurements and correlations of turbulent burning velocities over wide ranges of fuels and elevated pressures. *Proc. Combust. Inst.* **2013**, *34*, 1519–1526. [CrossRef]
23. Burke, E.M.; Güthe, F.; Monaghan, R.F.D. A comparison of turbulent flame speed correlations for hydrocarbon fuels at elevated pressures. In Proceedings of the ASME Turbo Expo 2016: Turbomachinery Technical Conference and Exposition, Seoul, Korea, 13–17 June 2016.
24. Muppala, S.P.R.; Aluri, N.K.; Dinkelacker, F.; Leipertz, A. Development of an algebraic reaction rate closure for the numerical calculation of turbulent premixed methane, ethylene, and propane/air flames for pressures up to 1.0 MPa. *Combust. Flame* **2005**, *140*, 257–266. [CrossRef]
25. Bradley, D.; Lawes, M.; Mansour, M.S. Correlation of turbulent burning velocities of ethanol–air, measured in a fan-stirred bomb up to 1.2 MPa. *Combust. Flame* **2011**, *158*, 123–138. [CrossRef]
26. Ratzke, A.; Schöffler, T.; Kuppa, K.; Dinkelacker, F. Validation of turbulent flame speed models for methane–air-mixtures at high pressure gas engine conditions. *Combust. Flame* **2015**, *162*, 2778–2787. [CrossRef]
27. Peng, M.W.; Shy, S.S.; Shiu, Y.W.; Liu, C.C. High pressure ignition kernel development and minimum ignition energy measurements in different regimes of premixed turbulent combustion. *Combust. Flame* **2013**, *160*, 1755–1766. [CrossRef]
28. Bradley, D.; Haq, M.Z.; Hicks, R.A.; Kitagawa, T.; Lawes, M.; Sheppard, C.G.W.; Woolley, R. Turbulent burning velocity, burned gas distribution, and associated flame surface definition. *Combust. Flame* **2003**, *133*, 415–430. [CrossRef]

29. Lipatnikov, A.N.; Li, W.Y.; Jiang, L.J.; Shy, S.S. Does density ratio significantly affect turbulent flame speed? *Flow Turbul. Combust.* **2017**, *98*, 1153–1172. [CrossRef] [PubMed]

30. Fogla, N.; Creta, F.; Matalon, M. The turbulent flame speed for low-to-moderate turbulence intensities: Hydrodynamic theory vs. experiments. *Combust. Flame* **2017**, *175*, 155–169. [CrossRef]

31. Peters, N. The turbulent burning velocity for large-scale and small-scale turbulence. *J. Fluid Mech.* **1999**, *384*, 107–132. [CrossRef]

32. Nakata, K.; Nogawa, S.; Takahashi, D.; Yoshihara, Y.; Kumagai, A.; Suzuki, T. Engine technologies for achieving 45% thermal efficiency of S.I. engine. *SAE Int. J. Eng.* **2015**, *9*, 179–192. [CrossRef]

33. Maruta, K.; Nakamura, H. Super lean-burn in SI engine and fundamental combustion studies. *J. Combust. SOC. Jpn.* **2016**, *58*, 9–19.

energies

MDPI

Article

Emissions from Solid Fuel Cook Stoves in the Himalayan Region

Jin Dang [1,*], Chaoliu Li [2], Jihua Li [3], Andy Dang [4,5], Qianggong Zhang [2], Pengfei Chen [6], Shichang Kang [2,6] and Derek Dunn-Rankin [1]

[1] Department of Mechanical and Aerospace Engineering, Henry Samueli School of Engineering, University of California, Irvine, CA 92697, USA; ddunnran@uci.edu
[2] Institute of Tibetan Plateau Research, Chinese Academy of Sciences, Beijing 100101, China; lichaoliu@itpcas.ac.cn (C.L.); Qianggong.zhang@itpcas.ac.cn (Q.Z.); shichang.kang@lzb.ac.cn (S.K.)
[3] Qujing Center for Disease Control and Prevention, Yunnan 655011, China; ynqj_cn@sina.com
[4] Program in Public Health, Susan and Henry Samueli College of Health Sciences, University of California, Irvine, CA 92697, USA; andyd@uci.edu
[5] Department of Epidemiology, School of Medicine, University of California, Irvine, CA 92697, USA
[6] State Key Laboratory of Cryosphere Science, Northwest Institute of Eco-Environment and Resources, Chinese Academy of Sciences, Lanzhou 730000, China; chenpengfeifeifei@126.com
* Correspondence: dangj1@uci.edu; Tel.: +1-949-385-1177

Received: 12 January 2019; Accepted: 12 March 2019; Published: 21 March 2019

Abstract: Solid fuel cooking stoves have been used as primary energy sources for residential cooking and heating activities throughout human history. It has been estimated that domestic combustion of solid fuels makes a considerable contribution to global greenhouse gas (GHG) and pollutant emissions. The majority of data collected from simulated tests in laboratories does not accurately reflect the performance of stoves in actual use. This study characterizes in-field emissions of fine particulate matter ($PM_{2.5}$), carbon dioxide (CO_2), carbon monoxide (CO), methane (CH_4), and total non-methane hydrocarbons (TNMHC) from residential cooking events with various fuel and stove types from villages in two provinces in China (Tibet and Yunnan) in the Himalayan area. Emissions of $PM_{2.5}$ and gas-phase pollutant concentrations were measured directly and corresponding emission factors calculated using the carbon balance approach. Real-time monitoring of indoor $PM_{2.5}$, CO_2, and CO concentrations was conducted simultaneously. Major factors responsible for emission variance among and between cooking stoves are discussed.

Keywords: solid fuel; cooking stove; field study

1. Introduction

Solid fuel cooking stoves continue to be used and relied upon in many parts of the world. There are more than two billion people using direct burning of solid fuel as their primary energy source [1,2], especially in developing countries where cooking stoves primarily burn biomass or coal. Furthermore, it has been estimated that worldwide domestic combustion of solid fuels from residential use and small=scale industry contribute approximately 34% of total black carbon (BC) emissions [3]. Biomass has been used directly as a fuel since the harnessing of fire by humans [4], and coal has been used since the second and third century of the Common Era [5]. Biomass fuels fall at the low end of the energy ladder and, consequently, require large volumes and mass relative to the energy delivered. As a result, they often produce a high level of combustion emissions. For household energy sources, the energy density ladder can be expressed as: dung < crop residues < wood < kerosene < gas < electricity [6]. Although switching to a higher energy ladder fuel or adopting new technology like gasification with co-generation provides a cleaner way to acquire energy [7], there are still large populations that use biomass and coal directly as fuel for cooking and heating. Coal has a high energy density, but also contains substantial levels of dangerous compounds, including sulfur and heavy metals. The wide-spread use of solid fuel due to human activity results not only in significant emission contribution to the atmosphere, but also negatively affects indoor air quality and public health.

Unlike other well-studied categories of combustion emission sources such as diesel engines [8–10], the emission inventory for the residential and small-scale industry sector is rarely investigated. In particular, depending on the type of fuel, emissions from solid fuel cooking stoves have a complicated make-up which includes well-mixed greenhouse gases (WMGHG) like carbon dioxide and methane, pollutants such as carbon monoxide, sulfur dioxide (mostly when coal is used as the fuel source), hydrocarbons, and particulate matter (PM), as well as small concentrations of volatile organic compounds [2]. The potential radiative forcing from these complex emissions is still unclear, especially for particulate matter [11]. This study aims to measure cooking stove emissions while they are in use to permit more accurate characterization of the potential local and global climate impact from domestic solid fuel combustion.

Studies of domestic solid fuel combustion emission have been underway for many years, but due to the limitation of technology deployment, the experimental study of biomass combustion emissions started only in the late 20th century [6,12–15]. With the help of statistical models, historical emissions data are available for major species including methane, carbon monoxide, nitrogen oxides, total and specialized non-methane volatile organic compounds (NMVOCs), ammonia, organic carbon, black carbon, and sulfur dioxide [16]. Unfortunately, the complexity and dispersivity of the emission sources means that the model study does not provide estimates with high accuracy and precision. In particular, several studies indicated that models consistently underestimate the carbon monoxide [17–21] and black carbon contributions resulting from biomass cooking stoves' use [22].

Controlled laboratory measurements of solid fuel cooking stoves have been made by many groups [23–26]. The widely-used testing protocol includes the water boiling test (WBT) and kitchen performance test (KPT). However, it has not been well-demonstrated that current testing protocols represent the actual everyday cooking and heating activities in homes, and there is still a lack of a confirmed explanation regarding the difference between laboratory and in-field measurements [27]. The actual emissions from household cooking stoves depend on several variables including: stove type, fuel type, food type, and the behavior of the cooks cooking the food. Laboratory experiments with uniformity and repeatability are not similar to everyday cooking and heating activities and, therefore, may not reflect the in-home conditions, nor the unavoidable variation of resident stove activities. This situation leads to highly uncertain in-field data [28].

Compared with laboratory experiment studies, in-field measurement has significant challenges in producing high-quality field data. For example, since most of the residents who use solid-fuel cooking stoves as their primary energy source live in rural areas, it is often hard to access these in-field sites. Furthermore, rural areas that rely on solid-fuel often have limited, or even no electrical power supply, which greatly restricts the measurement capabilities [29]. Moreover, taking measurements in homes is not as straightforward as doing so in a laboratory since the in-field environment is generally in an active family location. Local coordination plays an important role in this process.

2. Methodology

Emission factors of carbon dioxide (CO_2), carbon monoxide (CO), methane (CH_4), and fine particles ($PM_{2.5}$) were measured during normal daily activities using the carbon balance approach. Real-time monitoring of indoor $PM_{2.5}$, CO_2, and CO concentrations was conducted simultaneously.

2.1. In-Field Sampling

Globally, the emission to the environment from household cooking stoves' activity largely depends on the regional population density. The Himalaya Mountains, as one of the largest fresh water resources in the world, have a population of more than two billion people living on the rivers that originate from this source. The high population density in the nearby area leads to a significant contribution of emissions from domestic residential combustion. Hence, the field sites for this study were selected to be in two provinces of China (Tibet Autonomous Region and Yunnan Province) that are closest to the Himalaya Mountain range.

In general, each field campaign lasted for approximately two months with about 6 weeks of measurement time and 1 or 2 weeks for pre- and post- preparation. The basis for the selection of households included the ability to measure a variety of region-specific primary and secondary stove and fuel types. Depending on the given sites, there were also constraints and considerations taken into account regarding household selection. The campaigns took place during summer of 2012 and 2013 in Tibet and Yunnan, respectively. Thirty eight household samples were collected in the Tibet measurement and 40 for Yunnan. Figure 1 shows the geographical location of the field sites.

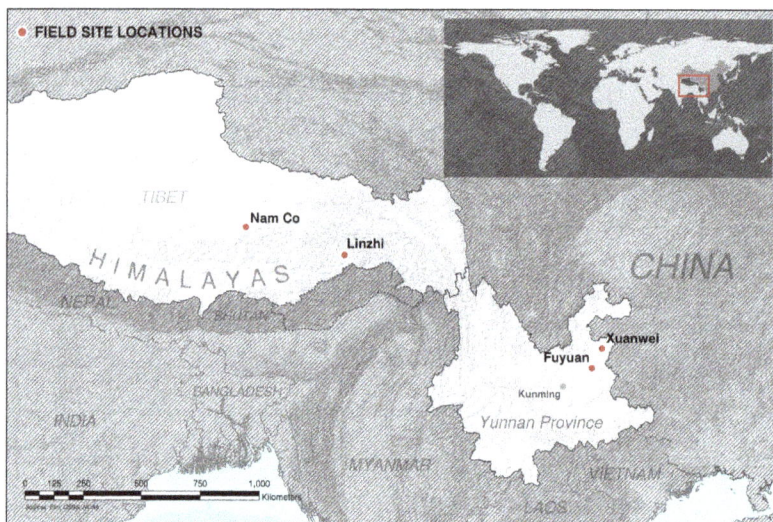

Figure 1. The geographical location of our field sites.

2.1.1. Tibet, China

The research in Tibet, China, was coordinated through the Institute of Tibetan Plateau Research, Chinese Academy of Sciences. Due to the limitation of the local environment, two regions, Nam Co and Linzhi, were chosen as the sites in which to conduct the in-field measurements. Figure 2 shows the household, stove, and fuel used in Tibet.

Nam Co has an extremely high elevation (approximately 4700 m). Due to the scarcity of plant and animal sources as biomass, local residents rely on yak dung as their primary fuel. Most of the local residents lead a traditional nomadic life, with a Tibetan tent as their shelter. The tourist industries being developed in the Nam Co area absorbed some residents to join the tourism business. As a result of their more settled living style, these people built fixed or semi-fixed households instead of traditional tents. There are two types of stoves being used in the Nam Co area: a traditional open fire stove and an improved chimney stove. The open fire stove is mostly used by nomadic residents, as it has better portability. All the fixed and semi-fixed households are now using an improved chimney stove. There was limited availability for household selection. As the population is very sparse in the Nam Co area, measurements were taken in whatever household could be found.

The Linzhi area has a much lower elevation (approximately 3300 m) compared to Nam Co. Lower elevation brings richer plant and animal sources. As a result, wood becomes the primary fuel for the Linzhi residents. The well-developed agriculture and tourist industry significantly improved the living condition of local residents. All the residents have well-built houses with well-designed chimney stoves (different from the ones in Nam Co). As there is a uniformity of households, stoves, and fuel type, the selection of participating households in the Linzhi area was mostly based on regional considerations. All the measured households in the Linzhi area were located in the two villages that are close to the Linzhi Research Station, Chinese Academy of Sciences.

Figure 2. The household, stove, and fuel in Tibet (**top left**: Tibetan tent in Nam Co; **top middle**: chimney stove; **top right**: household in Linzhi; **bottom left**: open fire stove in Nam Co; **bottom middle**: household in Nam Co; **bottom right**: yak dung).

The sampling system contained two sampling trains: one for cooking stove emission and the other for background monitoring. Starting with the probe, the emission sample traveled through a length (depending on the stove, typically around 2 m) of conductive tubing where it entered the sample train. The particle loss through 3 meters of this tubing was measured at about 2% by the University of Illinois Urbana-Champaign (UIUC) field sampling team [30]. A dilution pump was utilized to avoid instrument saturation and also to reflect natural dilution of chimney emissions. A cyclone separator cut off particles larger than 2.5 μm at a fixed flow rate (1.5 L/min). The first branch of the train collected elemental carbon (EC), organic carbon (OC), and gas samples. The EC/OC sample were collected with a 47-mm quartz filter. The gas sample was collected with a 200-L Kynar bag. The sampling duration generally started from breakfast cooking to the end of dinner cooking. However, Tibetan residences do not hold a regular cooking schedule as people from most of the other places due to their nomadic tradition. On the other hand, being constrained to this logistical limitation, in this case, the sampling event started around mid-morning and lasted approximately 7 h. The flow rate was constrained at 0.2 L/min with an SKC pocket sampling pump to collect all the gas with the 200-L bag. The second branch included one 37-mm PTFE filter and one 47-mm quartz filter to collect PM samples and EC/OC in the gas phase. The filtered gas traveled through a TSI Q-Trak 7575 CO/CO_2 monitor and Drager PAC 7000 monitor with a SO_2 sensor to acquire real-time CO, CO_2, and SO_2 information. The flow rate for the second branch was initially set at about 0.2 L/min. The third branch was for a DustTrak real-time aerosol monitor. The flow rate was set at 1.1 L/min to satisfy the 1.5-L/min requirement of the cyclone.

2.1.2. Yunnan, China

The study in Yunnan province was in cooperation with the Center for Disease Control and Prevention (CDC) of Qujing City. Yunnan is a rich coal province with mild climate, which brings an abundance of forestry and agricultural resources (elevation: approximately 2000 m). As a result, residents in Yunnan province have various fuels to choose from: coal, wood, corn, pine needles, and agricultural residue. However, people mostly use wood, corn, pine needles, or other biomass fuel to start the fire and then use coal to keep the stove burning (Figure 3). Hence, coal is considered as the primary fuel in Yunnan province, and covering all of the mainstream coal types (e.g., gas fat, smoky, coking, and smokeless) was the main consideration while selecting participating households.

The stove usage situation in Yunnan province is quite different from the other sites. As Yunnan province is relatively developed compared with Tibet, there are well-constructed electricity grids in this area, which provide local residents reliable and affordable power sources. As a result, many residents in the village switched to electric stoves (induction cooktop) for their primary cooking. Solid fuel stoves are still widely used for heating during the cold weather season (especially for the elderly) and preparing food for animals (electric stoves are not large enough for this task).

The solid fuel cooking stoves used in Yunnan generally are one of three different types: high stove, low stove, and portable stove (Figure 3). The high stove is a chimney stove, which is mainly used for cooking (before the electric stove became popular); the current low stove has a similar design to the high stove, but it sits lower to the ground level, which makes it a good floor heater. The portable stove is popular in villages, especially among seniors, as it is good for heating and easy to carry around.

Figure 3. The stove and fuel in Yunnan (**top left**: high stove; **top middle**: portable stove; **top right**: low stove, **bottom left**: coal, **bottom right**: corn).

2.2. The Carbon Balance Method

The measurement and analysis relied on the carbon balance method [31], which is commonly used for biomass combustion emission studies [32]. This method calculates the emission factor based on the carbon processed during fuel consumption and the ratio between pollutants in the exhaust gas [2,29]. In order to achieve a representative measurement, the sample is taken after the plume is well-mixed [23]. A prior study indicated that emissions take less than 2.5 s to reach phase equilibrium [33]. In our setup for open-fire stoves, the sample probe was located about 1 m above the stove, which left 3–4 s for the plume to mix before reaching the probe. For the chimney stove measurements, as the length of the chimney was mostly more than 2 m long and the flow inside the chimney was turbulent ($Re > 4000$) [23], we assumed the emissions were well-mixed within the chimneys, and we collected our sample from the chimney outlet.

2.3. Post-Measurement Analysis

The post-measurement analysis included gravimetric analysis for the $PM_{2.5}$ sample (PTFE filter) and gas chromatography analysis for the gas sample. The analysis of the EC/OC (quartz filter analysis) was conducted by collaborators in the UIUC group using a Sunset Laboratory OC/EC analyzer (thermal optical transmittance method) [34].

2.3.1. Gravimetric Analysis

Gravimetric analysis was applied to the PTFE filters collected from field measurements. Before heading to the field, the PTFE filters were weighed and sealed (pre-weights). A post-weight was conducted after the measurement in the field with the PM sample collected on the filter. With the recorded flow information and the weight difference between pre- and post-analysis, the $PM_{2.5}$ emission was calculated.

2.3.2. Gas Chromatography Analysis

Gas chromatography analysis was applied to the collected gas sample to investigate the concentration for interesting gas species, which include carbon dioxide, carbon monoxide, methane, and total hydrocarbon (THC). The detection of carbon dioxide and carbon monoxide were achieved with a flame ionization detector (FID) plus nickel catalyst methanizer (SRI Instruments, USA). The THC analysis was accomplished using a blank column plus FID. This approach takes advantage of the fact that the FID responds only to hydrocarbons [12]. External standardization was selected to calibrate the GC analysis. Calibration gases with known concentrations of target gas species were used to generate the calibration curve.

3. Results

The finalized database included emission factors for: carbon dioxide, carbon monoxide, methane, and $PM_{2.5}$ for each sample from the field sites. The emission factors were calculated with the carbon balance method. Modified combustion efficiency (MCE) was defined as the ratio between carbon in the form of carbon dioxide to that in the form of carbon dioxide plus that in the form of carbon monoxide. MCE is commonly used as an approximation of nominal combustion efficiency. Since most of the carbon emissions are in form of CO_2 and CO, MCE provides a robust approximation to the normalized combustion efficiency (NCE) [12].

$$MCE \equiv \frac{[CO_2]}{[CO_2] + [CO]} \tag{1}$$

3.1. Tibet

Table 1 provides the summary of the Tibet measurement. Thirty eight valid samples were collected: 28 samples from Nam Co and 12 from Linzhi. There were two small-scale industry measurements conducted in Nam Co. Both were small restaurants with exactly the same type of stoves used in local households.

The sparse population and limited transportation capability restricted the sampling time in each household. Tibet testing time was limited to around 6 h. Figure 4 shows a examples of the real-time pollutant concentration in Nam Co and Linzhi. The emission pattern from Tibet did not show obvious cooking events. This difference was attributed to the lifestyle of Tibetan residents and the desire for space heating for the stove. Particularly, in the Nam Co area, the nomadic life does not have a regular daily meal schedule (breakfast, lunch, snack, and dinner). As shown in Figure 4, the daytime stove activity in Tibet did not give obvious patterns representing regular cooking events. The cold weather in these high altitude regions encourages local residents to use cooking stoves for heating purposes, as well. The statistical summary for the Tibet measurements is shown in Table 2.

Table 1. Tibet household summary.

Location	Residence Type	Stove Type	Fuel Type	Measurement	# of Samples
Nam Co	Tent	Open fire	Yak dung	1-Day	4
	Tent	Chimney	Yak dung	1-Day	4
	Prefab house	Chimney	Yak dung	1-Day	12
	Stone house	Chimney	Yak dung	1-Day	6
	SSI	Chimney	Yak dung	1-Day	2
Linzhi	Garret	Chimney	Wood	3-Day	4
	Garret	Chimney	Wood	1-Day	8
Total					38

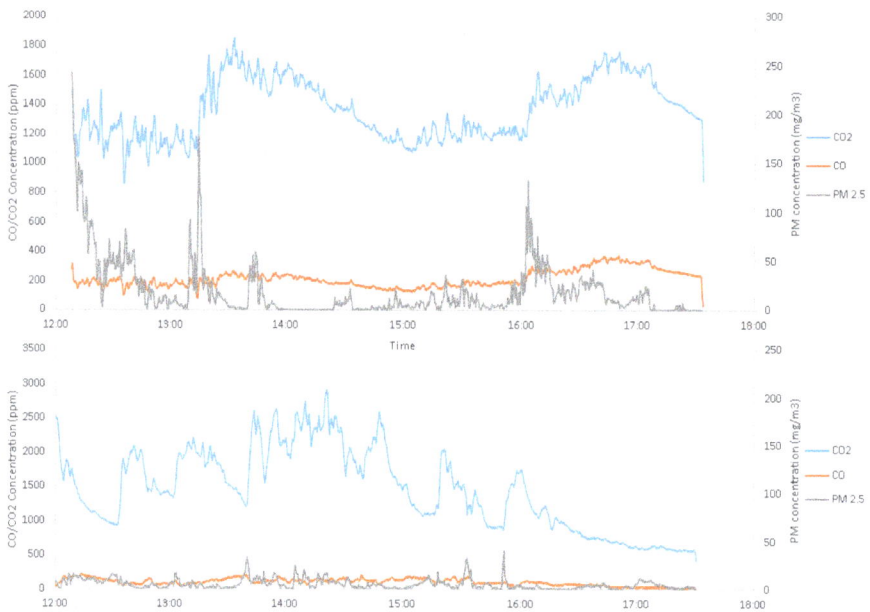

Figure 4. Typical real-time emission in Tibet, top: Nam Co; bottom: Linzhi.

Table 2. Statistical summary for the Tibet measurement.

	Sample Amount	MCE (%)	EF CO_2 (g/kg Fuel)	EF CO (g/kg Fuel)	EF $PM_{2.5}$ (g/kg Fuel)	EF CH_4 (g/kg Fuel)
Dung, open fire stove in tent	3	73.0 ± 7.5	1298.8 ± 148.3	303.1 ± 80.8	18.5 ± 10.2	32.4 ± 7
Dung, chimney stove in tent	2	90.2 ± 7.1	1590.4 ± 176.3	107.1 ± 76.4	26 ± 26	5.8 ± 1.6
Dung, chimney stove in house	15	91.4 ± 1.8	1632.4 ± 38.1	96 ± 20	14.7 ± 4.1	22 ± 4.6
Wood, chimney stove in house	15	84.0 ± 3.5	1282.9 ± 64.4	150.7 ± 31.5	18.6 ± 3.8	9.8 ± 1.3
All dung	22	88.6 ± 2.1	1579 ± 41.5	127.7 ± 23.2	16.3 ± 3.6	24.1 ± 3.8
All household	35	86.6 ± 2.0	1451.6 ± 44.9	137.8 ± 19.6	17.3 ± 2.7	16.7 ± 2.5
All SSI	2	89.1 ± 4.6	1587.5 ± 97.5	122.6 ± 51.5	16.1 ± 7.7	46.1 ± 5.7
Overall	37	86.7 ± 1.9	1459 ± 42.9	137 ± 18.6	17.2 ± 2.6	18.3 ± 2.6

3.2. Yunnan

Table 3 shows the summary of the Yunnan measurements. The Yunnan measurement set contained 40 valid samples from two counties with 10 villages. The household types in Yunnan are uniform, while stove type varies.

Table 3. Yunnan household summary.

Location	Residence Type	Stove Type	Fuel Type	Measurement	# of Samples
Fuyuan	House	High stove	Coal	1-Day	13
	House	High stove	Coal	3-Day	2
	House	Portable stove	Coal	1-Day	5
	House	Portable stove	Coal	3-Day	1
	House	Low stove	Coal	1-Day	1
Xuanwei	House	High stove	Coal	1-Day	7
	House	High stove	Coal	3-Day	1
	House	Portable stove	Coal	1-Day	8
	House	Portable stove	Coal	3-Day	1
	House	Low stove	Coal	1-Day	1
Total					40

Figure 5 shows an example of the pollutant concentration throughout the day. The two major peak emission event groups indicated the traditional breakfast and lunch event in a Chinese village. The dinner cooking event, which usually happens at approximately 18:00, could not be monitored due to local collaborator unavailability. The smaller peak at around 15:00 was normally caused by water boiling or a snack event. As corn is often used as the stove starter, the moisture content in it generated a large amount of smoke at the starting stage of each cooking event; this process showed a strong $PM_{2.5}$ peak concentration at the beginning of every cooking event, while the concentration of CO and CO_2 did not have the same behavior.

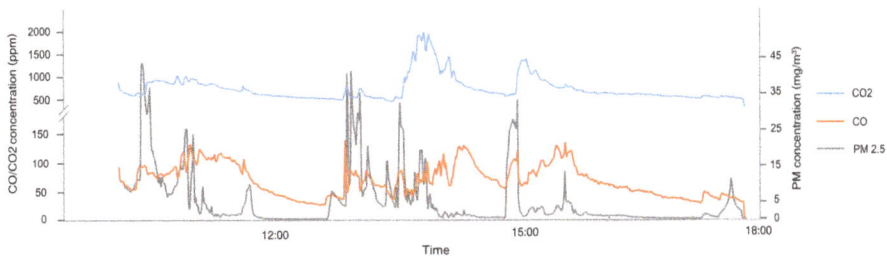

Figure 5. Typical real-time emission pattern for CO_2, CO, and $PM_{2.5}$ in Yunnan.

Table 4 is the summary for the Yunnan dataset. As previously mentioned, mixed fuel usage is very popular in Yunnan. The native definition of the emission factor determined that it was very sensitive to fuel consumption [15]. Thus, it is important to separate when using agricultural residue as a lighter only and using coal as the major energy source and using both agricultural residue and coal as the energy source. Based on observations in the field, the threshold value of 2 kg of agricultural residue consumption was selected as the criteria of whether a household was using agricultural residue as a lighter only. In the case of more than 2 kg of agricultural residue consumed, the emission factors were calculated based on the weight of total fuel (coal plus agricultural residue). In the other case, only coal consumption was considered in the emission factor calculation.

Table 4. Statistical summary for the Yunnan measurement.

	Sample Amount	MCE (%)	EF CO_2 (g/kg Fuel)	EF CO (g/kg Fuel)	EF $PM_{2.5}$ (g/kg Fuel)	EF CH_4 (g/kg Fuel)
High stove in Fuyuan	15	82.7 ± 2.7	1528.366 ± 174.458	194.982 ± 30.674	16.585 ± 3.49	83.584 ± 17.85
Portable stove in Fuyuan	6	87.1 ± 3.1	1973.451 ± 217.36	199.102 ± 60.626	12.052 ± 2.311	106.308 ± 24.3
Low stove in Fuyuan	1	85.2	1379.019	152.8224	32.09369	89.74726
High stove in Xuanwei	9	85.8 ± 1.4	1573.745 ± 251.195	172.03 ± 34.978	21.757 ± 8.609	61.563 ± 11.245
Portable stove in Xuanwei	11	88.2 ± 0.7	1457.356 ± 269.818	125.194 ± 25.393	13.323 ± 4.251	88.588 ± 20.922
Low stove in Xuanwei	1	89.7	2437.381	178.345	2.360746	54.69838
Overall Fuyuan	22	84.0 ± 2.1	1642.964 ± 137.47	194.189 ± 25.892	16.054 ± 2.58	90.061 ± 13.71
Overall Xuanwei	21	87.2 ± 0.7	1553.905 ± 178.872	147.798 ± 20.169	16.415 ± 4.332	75.392 ± 12.083
Overall high stove	24	83.8 ± 1.8	1545.383 ± 140.819	186.375 ± 22.867	18.525 ± 3.817	75.326 ± 11.943
Overall portable stove	17	87.8 ± 1.1	1639.507 ± 196.098	151.279 ± 27.279	12.874 ± 2.814	94.842 ± 15.704
Overall low stove	2	87.4 ± 2.3	1908.2 ± 529.181	165.584 ± 12.761	17.227 ± 14.866	72.223 ± 17.524
Overall	43	85.6 ± 1.1	1599.47 ± 111.006	171.533 ± 16.7	16.23 ± 2.463	82.897 ± 9.128

In order to explore the effect from mixing agricultural residue (mostly corn cob) together with coal, the correlation of modified combustion efficiency with different fuel mixing factors is plotted in Figure 6. The mixing ratio here is defined as the fraction of coal consumption (kg) in the total fuel consumption (kg). Modified combustion efficiency, which is essentially a comparison of how much carbon is emitted in CO_2 and CO form, was used as the indicator because it is estimated independently of fuel information. The result in Figure 6 shows that MCE distributed evenly through the whole fuel mixing ratio range, which implies that the mixing ratio did not have a significant effect on the stove performance.

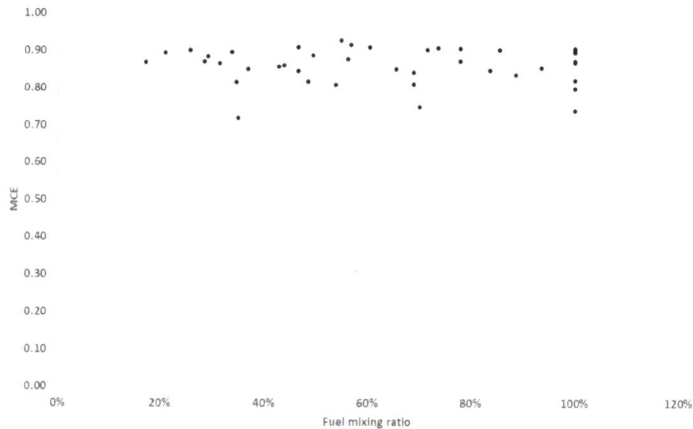

Figure 6. MCE at different fuel mixing ratios.

4. Discussion

4.1. Efficiency and Emissions

Figure 7 compares the modified combustion efficiency for wood burning cooking stoves between this study (with uncertainties) and several previous works, including both the water boiling test and the measurement of actual stove usage. The MCE measured from actual stove use in the field was consistently lower when compared with those measured from standard water boiling tests. A comparison for the CO emission factor (Figure 8) gave similar results. The measurements on actual stove use in homes gave significantly higher CO emission and larger uncertainty than the WBTs.

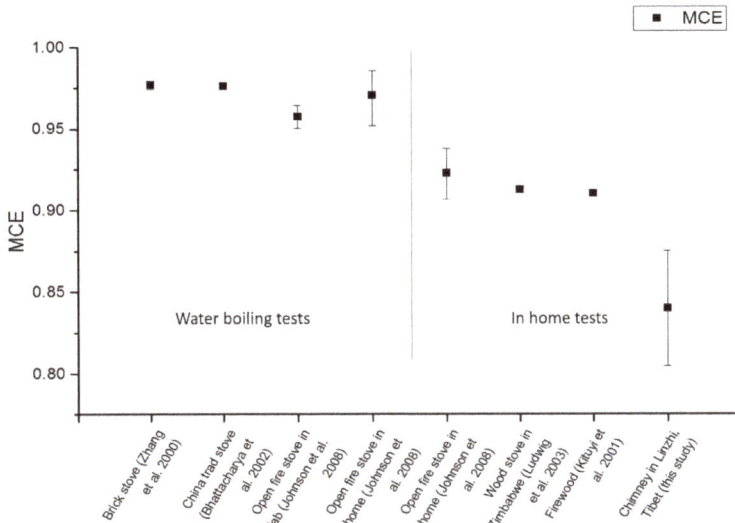

Figure 7. The MCE comparison.

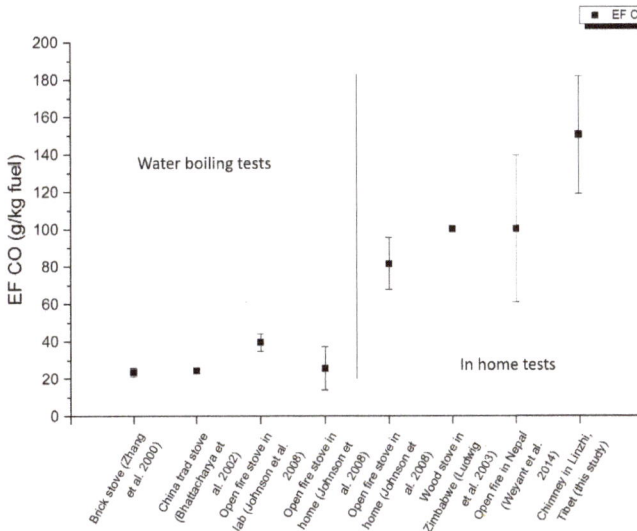

Figure 8. The CO emission factor comparison.

Figure 9 compares the emission parameters from different regions with different fuels within this study. The MCE results from all regions with all different fuels showed a similar average value.

The CO_2 emission factor is a good indicator for carbon emission, as most emissions come out in the form of CO_2. As it has been well-addressed, fuel makes a big difference in the carbon emission. One major reason for this is the carbon density in fuel, which directly affects the emission factor of carbon compounds. For example, trunk woods normally have a carbon content of about 50%. The carbon content for coal can be over 90% [35]. Comparing the CO_2 emission factor for the wood stoves (Tibet) and coal stoves (Yunnan), the carbon emission from coal stoves was significantly higher (Figure 9).

CO and $PM_{2.5}$ are the major products of incomplete combustion, and they are closely related to indoor air quality and residents' health. One challenge for CO and $PM_{2.5}$ measurements is the associated uncertainties, especially for $PM_{2.5}$. Dung is generally considered as a more "dirty" fuel. However, the yak dung stove in Nam Co, Tibet, gave lower CO and $PM_{2.5}$ than the wood stoves in Linzhi, Tibet. As previously discussed, the stoves in Nam Co, Tibet, due to the harsh environment (high altitude, cold), are partially used as heating stoves. This difference on "how the stove is used" may explain this unusual behavior. Among all the field sites in this study, Yunnan coal stoves had the highest CO and $PM_{2.5}$ emission factors. According to the local CDC, the field sites in Yunnan Province, Fuyuan and Xuanwei Counties, all have a high occurrence of lung cancer.

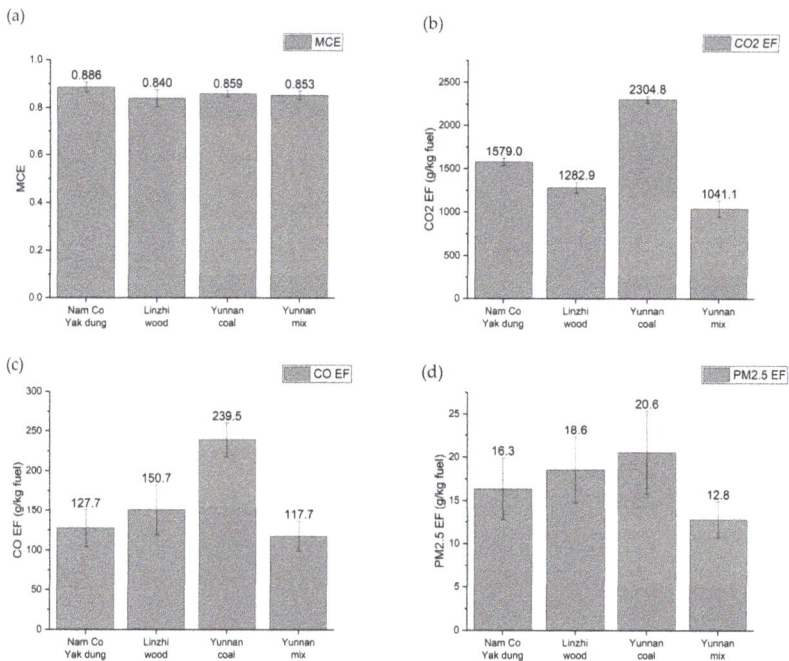

Figure 9. Comparison of emission parameters between sites and fuels: (**a**) Modified combustion efficiency (MCE); (**b**) Emission factor for CO_2; (**c**) Emission factor for CO; (**d**) Emission factor for $PM_{2.5}$.

4.2. Carbon Particulate Emission

EC/OC particulate measurements are inherently noisier than gas measurements and total particulate measurements because they are often less uniformly mixed at the sampling zone (the complete mixing into the atmosphere occurs over a longer time). Sample numbers of EC and OC were also much smaller than for the other species. Nevertheless, the EC/OC data are a critical component of the uncertainty in climate forcing, so it is important to include even the limited validated data obtained. The summary of EC and OC measurement is shown in Table 5.

Table 5. Summary of elemental carbon and organic carbon results.

Location	Fuel	Sample Amount	EF EC (g/kg Fuel)	EF OC (g/kg Fuel)
Tibet, China	Yak Dung	10	0.25 ± 0.05	15.41 ± 2.54
Tibet, China	Wood	2	0.11 ± 0.05	16.03 ± 14.48
Yunnan, China	Coal	16	1.46 ± 0.47	10.09 ± 2.71
Yunnan, China	Mix	18	0.51 ± 0.17	7.02 ± 1.26

Figure 10 shows a comparison of EC and OC emission factors through all field sites and fuels. The coal stoves in Yunnan, China, produced the highest elemental carbon emissions, even considering its high uncertainty, while other stoves burning agricultural-based fuel had a similar result. A plot of the EC and OC ratio for the different regions and fuel combinations is shown in Figure 11, as this ratio is usually of more interest.

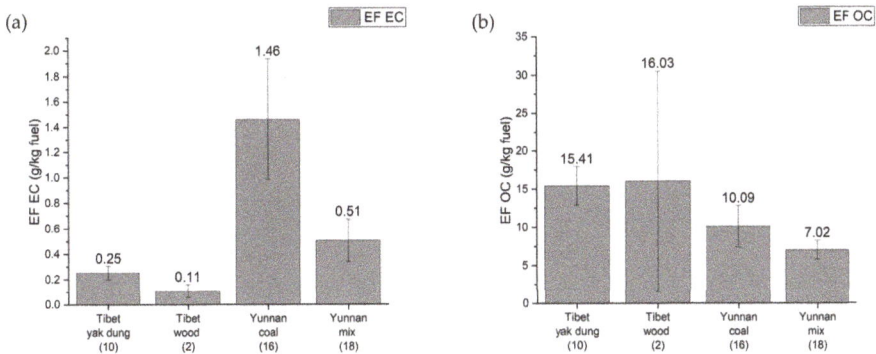

Figure 10. Comparison of carbon particulate emission factor between sites and fuels: (**a**) Elemental carbon; (**b**) Organic carbon.

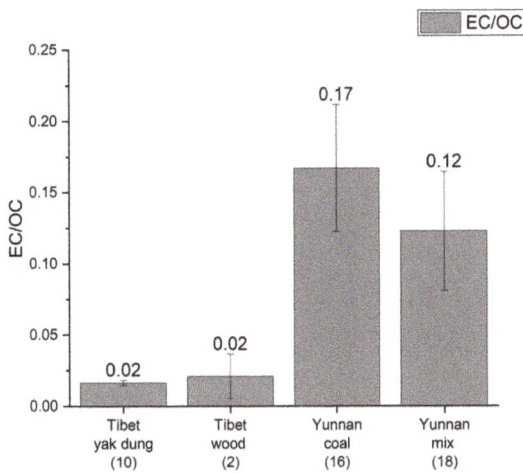

Figure 11. Comparison of the EC/OC ratios between sites and fuels.

Elemental carbon emission relates closely to the carbon content in the fuel: a higher carbon fuel such as coal produces more EC during the combustion process. From the perspective of combustion chemistry, the formation of soot, which is mostly elemental carbon, is largely related to the production of acetylene (C_2H_2). The agricultural-based fuels (wood, dung, agricultural residue, etc.) essentially consist of lingo-cellulosic materials, with hydrogen, carbon, and oxygen the dominate elements. In the combustion process, before soot (EC) is produced, the long carbohydrate polymers (larger C number) break up into shorter carbohydrates (smaller C number) until acetylene. For a stove that does not burn fuel completely (which is common for cooking stoves), the decomposition reactions for some molecules can stop before reaching the acetylene stage. In this case, organic carbon is emitted. Elemental carbon, on the other hand, means that reactions reach the key soot precursor acetylene, which represents more complete combustion as compared to high OC emission combustion.

5. Conclusions

A series of in-field emission measurements for solid fuel cooking stoves has been conducted. Emission factors for major compounds, including CO_2, CO, $PM_{2.5}$, CH_4, and elemental and organic carbon, with their statistical uncertainties were calculated using the carbon balance method. The acquired emission database fills in the inventory for residential cooking stoves' emission in the corresponding regions. These unique data from in-field measurements showed the real variability of cooking stove use and caution against simple methods to incorporate cooking stove emissions in global climate models. The results show clearly that in average, real-life cooking, stove use emits more than occurs under laboratory and controlled conditions.

The result from these field measurements show that fuel type is a critical variable in cooking stove performance. However, the effect of stove activity, such as the fluctuation during the combustion process, and the start and stop stages, should be considered in modeling and designing controlled laboratory experiments.

Author Contributions: Methodology, J.D.; formal analysis, J.D.; investigation, J.D., A.D., C.L., Q.Z., P.C., S.K., and J.L.; resources, J.D., A.D., C.L., Q.Z., P.C., S.K., and J.L.; data curation, J.D., A.D., and J.L.; writing, original draft preparation, J.D.; writing, review and editing, J.D., A.D., and D.D.-R.; visualization, J.D., and A.D.; supervision, D.D.-R.; project administration, A.D.

Funding: This research has been supported by a grant from the U.S. Environmental Protection Agency (EPA)'s Science to Achieve Results (STAR) through its Office of Research and Development in the research described here under Grant Number 83503601. It has not been subject to Agency review and therefore does not necessarily reflect the views of the U.S. EPA. The contents are solely the responsibility of the authors. No official endorsement should be inferred. The mention of trade names or commercial products in the publication does not constitute endorsement or recommendation for use.

Acknowledgments: We would like to acknowledge our collaborators from the University of Illinois at Urbana-Champaign, Tami Bond, Cheryl L. Weyant, and Ryan Thompson, and their research support staff for their significant contributions in the implementation, execution, and sample analyses. Many thanks to Rufus Edwards for his supervision and funding acquisition. Special acknowledgments to Kirk Smith for his original conceptualization and research design that laid the foundation and groundwork for this study. Additional acknowledgments to Qing Lan, Nathaniel Rothman, and Wei Hu for facilitating field measurements in Yunnan Province, China. We would also like to thank the following collaborators: the Institute of Tibetan Plateau Research, Chinese Academy of Sciences, and the Center for Disease Control and Prevention (CDC) of Qujing City for their assistance in the field. Special thanks to Allison Mok, Harman Chauhan, Vy Pham, Stephanie Fong, Kunaal Kapoor, and Scott Ondap for their contributions.

Conflicts of Interest: The authors declare no conflict of interest. The funders had no role in the design of the study; in the collection, analyses, or interpretation of data; in the writing of the manuscript; nor in the decision to publish the results.

References

1. Bruce, N.; Rehfuess, E.; Mehta, S.; Hutton, G.; Smith, K. Indoor air pollution. In *Disease Control Priorities in Developing Countries*, 2nd ed.; Jamison, D.T., Breman, J.G., Measham, A.R., Alleyne, G., Claeson, M., Evans, D.B., Jha, P., Mills, A., Musgrove, P., Eds.; World Bank: Washington, DC, USA, 2006.
2. Jetter, J.J.; Kariher, P. Solid-fuel household cooking stoves: Characterization of performance and emissions. *Biomass Bioenergy* **2009**, *33*, 294–305. [CrossRef]
3. Bond, T.C.; Bhardwaj, E.; Dong, R.; Jogani, R.; Jung, S.; Roden, C.; Streets, D.G.; Trautmann, N.M. Historical emissions of black and organic carbon aerosol from energy-related combustion, 1850–2000. *Glob. Biogeochem. Cycles* **2007**, *21*, GB2018. [CrossRef]
4. Turns, S.R. *An Introduction to Combustion, Concepts and Applications*; McGraw Hill: New York, NY, USA, 2012.
5. *DOE—Fossil Energy: A Brief History of Coal Use in the United States*; Department of Energy: Washington, DC, USA, 2019.
6. Smith, K.R.; Apte, M.G.; Yuqing, M.; Wongsekiarttirat, W.; Kulkarni, A. Air pollution and the energy ladder in asian cities. *Energy* **1994**, *19*, 587–600. [CrossRef]
7. Ahrenfeldt, J.; Thomsen, T.P.; Henriksen, U.; Clausen, L.R. Biomass gasification cogeneration—A review of state of the art technology and near future perspectives. *Appl. Therm. Eng.* **2013**, *50*, 1407–1417. [CrossRef]
8. Subramanian, R.; Winijkul, E.; Bond, T.C.; Thiansathit, W.; Oanh, N.T.K.; Paw-Armart, I.; Duleep, K. Climate-relevant properties of diesel particulate emissions: results from a piggyback study in Bangkok, Thailand. *Environ. Sci. Technol.* **2009**, *43*, 4213–4218. [CrossRef] [PubMed]
9. Maricq, M.M.; Podsiadlik, D.H.; Chase, R.E. Examination of the size-resolved and transient nature of motor vehicle particle emissions. *Environ. Sci. Technol.* **1999**, *33*, 1618–1626. [CrossRef]
10. Yan, F.; Winijkul, E.; Jung, S.; Bond, T.C.; Streets, D.G. Global emission projections of particulate matter (PM): I. Exhaust emissions from on-road vehicles. *Atmos. Environ.* **2011**, *45*, 4830–4844. [CrossRef]
11. Bond, T.C.; Doherty, S.J.; Fahey, D.W.; Forster, P.M.; Berntsen, T.; DeAngelo, B.J.; Flanner, M.G.; Ghan, S.; Kärcher, B.; Koch, D.; et al. Bounding the role of black carbon in the climate system: A scientific assessment. *J. Geophys. Res. Atmos.* **2013**, *118*, 5380–5552. [CrossRef]
12. Johnson, M.; Edwards, R.; Alatorre Frenk, C.; Masera, O. In-field greenhouse gas emissions from cookstoves in rural Mexican households. *Atmos. Environ.* **2008**, *42*, 1206–1222. [CrossRef]
13. Zhang, J.; Smith, K.R. Hydrocarbon emissions and health risks from cookstoves in developing countries. *J. Expo. Anal. Environ. Epidemiol.* **1995**, *6*, 147–161.
14. Zhang, J.; Smith, K.R.; Uma, R.; Ma, Y.; Kishore, V.V.N.; Lata, K.; Khalil, M.A.K.; Rasmussen, R.A.; Thorneloe, S.T. Carbon monoxide from cookstoves in developing countries: 1. Emission factors. *Chemosph. Glob. Chang. Sci.* **1999**, *1*, 353–366. [CrossRef]
15. Zhang, J.; Smith, K.R. Emissions of carbonyl compounds from various cookstoves in China. *Environ. Sci. Technol.* **1999**, *33*, 2311–2320. [CrossRef]
16. Lamarque, J.F.; Bond, T.C.; Eyring, V.; Granier, C.; Heil, A.; Klimont, Z.; Lee, D.; Liousse, C.; Mieville, A.; Owen, B.; et al. Historical (1850–2000) gridded anthropogenic and biomass burning emissions of reactive gases and aerosols: Methodology and application. *Atmos. Chem. Phys.* **2010**, *10*, 7017–7039. [CrossRef]
17. Bergamaschi, P.; Hein, R.; Heimann, M.; Crutzen, P.J. Inverse modeling of the global CO cycle: 1. Inversion of CO mixing ratios. *J. Geophys. Res. Atmos.* **2000**, *105*, 1909–1927. [CrossRef]
18. Kasibhatla, P.; Arellano, A.; Logan, J.A.; Palmer, P.I.; Novelli, P. Top-down estimate of a large source of atmospheric carbon monoxide associated with fuel combustion in Asia. *Geophys. Res. Lett.* **2002**, *29*, 6. [CrossRef]
19. Gros, V.; Williams, J.; Lawrence, M.; Von Kuhlmann, R.; Van Aardenne, J.; Atlas, E.; Chuck, A.; Edwards, D.; Stroud, V.; Krol, M. Tracing the origin and ages of interlaced atmospheric pollution events over the tropical Atlantic Ocean with in situ measurements, satellites, trajectories, emission inventories, and global models. *J. Geophys. Res. Atmos.* **2004**, *109*. [CrossRef]
20. Streets, D.G.; Zhang, Q.; Wang, L.; He, K.; Hao, J.; Wu, Y.; Tang, Y.; Carmichael, G.R. Revisiting China's CO emissions after the transport and chemical evolution over the Pacific (TRACE-P) mission: Synthesis of inventories, atmospheric modeling, and observations. *J. Geophys. Res. Atmos.* **2006**, *111*. [CrossRef]

21. Venkataraman, C.; Habib, G.; Kadamba, D.; Shrivastava, M.; Leon, J.F.; Crouzille, B.; Boucher, O.; Streets, D. Emissions from open biomass burning in India: Integrating the inventory approach with high-resolution Moderate Resolution Imaging Spectroradiometer (MODIS) active-fire and land cover data. *Glob. Biogeochem. Cycles* **2006**, *20*. [CrossRef]

22. Koch, D.; Bond, T.C.; Streets, D.; Unger, N.; Van der Werf, G.R. Global impacts of aerosols from particular source regions and sectors. *J. Geophys. Res. Atmos.* **2007**, *112*. [CrossRef]

23. Roden, C.A.; Bond, T.C.; Conway, S.; Osorto Pinel, A.B.; MacCarty, N.; Still, D. Laboratory and field investigations of particulate and carbon monoxide emissions from traditional and improved cookstoves. *Atmos. Environ.* **2009**, *43*, 1170–1181. [CrossRef]

24. Zhang, J.; Smith, K.R.; Ma, Y.; Ye, S.; Jiang, F.; Qi, W.; Liu, P.; Khalil, M.A.K.; Rasmussen, R.A.; Thorneloe, S.A. Greenhouse gases and other airborne pollutants from household stoves in China: A database for emission factors. *Atmos. Environ.* **2000**, *34*, 4537–4549. [CrossRef]

25. Smith, K.R.; Khalil, M.a.K.; Rasmussen, R.A.; Thorneloe, S.A.; Manegdeg, F.; Apte, M. Greenhouse gases from biomass and fossil fuel stoves in developing countries: A Manila pilot study. *Chemosphere* **1993**, *26*, 479–505. [CrossRef]

26. Bhattacharya, S.C.; Albina, D.O.; Abdul Salam, P. Emission factors of wood and charcoal-fired cookstoves. *Biomass Bioenergy* **2002**, *23*, 453–469. [CrossRef]

27. Taylor, R.P. The Uses of Laboratory Testing of Biomass Cookstoves and the Shortcomings of the Dominant U.S. Protocol. Ph.D. Thesis, Iowa State University, Ames, IA, USA, 2009.

28. Bond, T.C.; Streets, D.G.; Yarber, K.F.; Nelson, S.M.; Woo, J.H.; Klimont, Z. A technology-based global inventory of black and organic carbon emissions from combustion. *J. Geophys. Res. Atmos.* **2004**, *109*, D14203. [CrossRef]

29. Roden, C.A.; Bond, T.C.; Conway, S.; Pinel, A.B.O. Emission factors and real-time optical properties of particles emitted from traditional wood burning cookstoves. *Environ. Sci. Technol.* **2006**, *40*, 6750–6757. [CrossRef] [PubMed]

30. Bond, T.C.; Anderson, T.L.; Campbell, D. Calibration and intercomparison of filter-based measurements of visible light absorption by aerosols. *Aerosol Sci. Technol.* **1999**, *30*, 582–600. [CrossRef]

31. Crutzen, P.J.; Heidt, L.E.; Krasnec, J.P.; Pollock, W.H.; Seiler, W. Biomass burning as a source of atmospheric gases CO, H_2, N_2O, NO, CH_3Cl and COS. *Nature* **1979**, *282*, 253–256. [CrossRef]

32. Ward, D.E.; Hao, W.M.; Susott, R.A.; Babbitt, R.E.; Shea, R.W.; Kauffman, J.B.; Justice, C.O. Effect of fuel composition on combustion efficiency and emission factors for African savanna ecosystems. *J. Geophys. Res. Atmos.* **1996**, *101*, 23569–23576. [CrossRef]

33. Lipsky, E.M.; Robinson, A.L. Design and evaluation of a portable dilution sampling system for measuring fine particle emissions. *Aerosol Sci. Technol.* **2005**, *39*, 542–553. [CrossRef]

34. *NIOSH Manual of Analytical Methods (NMAM)*, 4th ed.; Third Supplement; NIOSH: Washington, DC, USA, 2018. [CrossRef]

35. Gaur, S.; Reed, T.B. *Thermal Data for Natural and Synthetic Fuels*; CRC Press: Boca Raton, FL, USA, 1998.

Article

Characterization of Performance of Short Stroke Engines with Valve Timing for Blended Bioethanol Internal Combustion

Kun-Ho Chen and Yei-Chin Chao *

Department of Aeronautics and Astronautics, National Cheng Kung University, 701 Tainan, Taiwan;
chenkh1979@gmail.com
* Correspondence: ycchao@mail.ncku.edu.tw; Tel.: +886-6-2757575 (ext. 63690)

Received: 19 January 2019; Accepted: 21 February 2019; Published: 25 February 2019

Abstract: The present study provides a feasible strategy for minimizing automotive CO_2 emissions by coupling the principle of the Atkinson cycle with the use of bioethanol fuel. Motor cycles and scooters have a stroke to bore ratio of less than unity, which allows higher speeds. The expansion to compression ratio (ECR) of these engines can be altered by tuning the opening time of the intake and exhaust valves. The effect of ECR on fuel consumption and the feasibility of ethanol fuels are still not clear, especially for short stroke engines. Hence, in this study, the valve timing of a short stroke engine was tuned in order to explore potential bioethanol applications. The effect of valve timing on engine performance was theoretically and experimentally investigated. In addition, the application of ethanol/gasoline blended fuels, E3, E20, E50, and E85, were examined. The results show that consumption, as well as engine performance of short stroke motorcycle engines, can be improved by correctly setting the valve controls. In addition, ethanol/gasoline blended fuel can be used up to a composition of 20% without engine modification. The ignition time needs to be adjusted in fuel with higher compositions of blended ethanol. The fuel economy of a short stroke engine cannot be sharply improved using an Atkinson cycle, but CO_2 emissions can be reduced using ethanol/gasoline blended fuel. The present study demonstrates the effect of ECR on the performance of short stroke engines, and explores the feasibility of applying ethanol/gasoline blended fuel to it.

Keywords: short stroke engine; bioethanol; Atkinson cycle

1. Introduction

By the end of 2015, the number of registered motorcycles in Taiwan totaled 21.4 million [1], tantamount to the overall population. In recent years, motorcycles and scooters have played an increasingly important role in transportation, in Taiwan, India, and other Southeast Asian countries. In order to reach the CO_2 emissions targets that have been widely agreed upon by industrialized nations, fossil fuel emissions caused by automobiles must be reduced. These global environmental issues and the depletion of fossil fuels can be partially addressed by increasing the use of biofuels and enhancing engine efficiency. The Atkinson cycle, which was proposed by James Atkinson [2] in 1882, has been proven to reduce fuel consumption through improved efficiency. The original Atkinson engine used a unique crankshaft that allowed the intake, compression, power, and exhaust strokes to occur during one turn of the crankshaft. The basic principle is that because the power (expansion) stroke is longer than the compression stroke, the efficiency is greater than that of an Otto-cycle engine. As shown in Figure 1, in modern applications, the engine achieves higher efficiency using a variable valve timing system, which achieves a power stroke that is longer than the compression stroke [3]. In short, the expansion to compression ratio (ECR) is over 1.0. The Atkinson cycle engine has been studied for decades with particular focus on variable valve control [4], performance analysis [5–7], and integrated applications [8]. Moreover, the concept of the Atkinson cycle has also been used to reduce emissions in diesel engines [9]. Atkinson cycle engines have been used in modern vehicles. e.g., the Toyota Prius. The Prius is equipped with a 1NZ-FXE engine which is a long stroke engine and suitable for the Atkinson cycle.

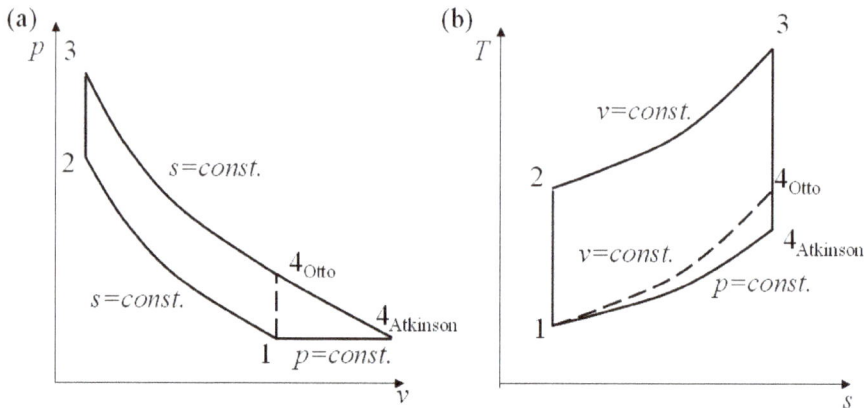

Figure 1. p-v and T-s diagram of Otto cycle and Atkinson cycle.

Alternatively, biomass has been proposed as a potential clean and renewable energy source [10]. Plants capture both solar energy and carbon dioxide to form plant body materials. In modern automotive applications, it is impractical to directly use traditional forms of biomass, e.g., dry wood or grain. Biomass needs to be transformed into gas or liquid form through chemical, thermochemical, or fermentation processes. In gasoline engines, ethanol can be used not only as an alternative fuel [11], but also as the oxygenated additive for gasoline to increase the octane number and to control emissions [12]. Moreover, greenhouse gas emissions can be reduced by using ethanol gasoline blends [13]. The performance of ethanol in IC engines has been studied for several decades [14–16], and the results show that the brake thermal efficiency (BTE), volumetric efficiency and brake mean effective pressure (BMEP) are all higher with ethanol. During 2008, the growth of the corn ethanol industry in the US developed for transportation fuel was blamed for a spike in global food prices [17]. The recent development of cellulosic ethanol [18,19], may allow bioethanol fuels to still play a significant role in automotive fuel in the near future [20].

The present study provides a feasible strategy for minimizing automotive CO_2 emissions by coupling the principle of the Atkinson cycle (Show in Figure 1) with the use of bioethanol fuel. Note that the Otto cycles and scooters have a stroke to bore ratio of less than unity, which allows higher speeds. The ECR of the engine can be altered by tuning the opening time of the intake and exhaust valves. The effect of ECR on fuel consumption and the feasibility of ethanol fuels are still not clear, especially for short stroke engines. Hence, a single cylinder, spark ignition, high speed engine is chosen as a platform for further examination. Briefly, there are four major objectives in the present study:

1. To theoretically define the proper intake and exhaust valve timing. The original engine performance is also calculated as the baseline.
2. To compare the calculated results with the measured data for the purpose of verification.
3. To identify the effect of the Atkinson cycle on the brake specific fuel consumption (BSFC) power output, torque, and emissions for short stroke engines.
4. To experimentally analyze engine performance with the proper intake and outtake valve timing, and to delineate the BSFC, power output, torque, and emissions for different ethanol blended fuels.

2. Methodology

The engine used for the present study was a single cylinder KYMCO 500CC engine with spark ignition, port injection, and which was electronically controlled with dual overhead cam. The specifications of the engine, as well as the valve timing controls are listed in Table 1. The engine was coupled to an API FR50 model eddy current dynamometer for power output, torque, and engine speed measurements. The emissions, which include O_2, HC, CO, CO_2, and NO, were measured using Horiba MEXA-584L (HORIBA, Ltd., Kyoto, Japan). Note that the flue gas was sampled prior to entering the catalytic reactor. In addition, an air to fuel ratio (AFR) sensor was also installed to monitor the exhaust gas. The quantities of air and fuel were measured by the air mass flowmeter (HFM-200 LFE, Teledyne Technologies Incorporated, Thousand Oaks, CA, USA) and fuel mass meter (F C MASSFLO 4100, Siemens AG), respectively. Similar to the pollution emission measurement for industry application, the measured pollution emissions were normalized to a specific O_2 concentration [21]. The emissions of the present study were corrected to 3% O_2. The method for transferring measured data to a standard basis is given by Equation (1) [22]

$$x_{corr} = x_{meas} \frac{20.95 - O_{2,basis}}{20.95 - O_{2,meas}}. \tag{1}$$

Table 1. Test engine specification.

Design	Specification
Engine type	Spark ignition engine
Cooling	Liquid cooled
no. of cylinders	1
Configuration	Port fuel injection
Bore	92 mm
Stroke	75 mm
Compression ratio	10.6:1
displacement	498.5 cc
Cylinder head	4 valve, DOHC

To calculate the heat release rate and total heat release rate, cylinder pressure was measured using a piezoelectric sensor D322D6-SO (Optrand, Inc., Plymouth, MI, USA). The shaft of the engine was equipped with an angular encoder to provide a reference for electronic control. The engine was fully controlled and monitored using a lab-made control system based on NI (National Instruments Corporation, Austin, TX, USA) real-time hardware (CompactRIO) with LabVIEW interface. The signals from the dynamometer, mass flow meter, temperature and pressure sensors, and gas analyzer were also logged, compiled, and stored in the PC for further analysis. The signals were collected for 60 s as the engine rpm reached steady, and the sampling rate was 20 Hz. The standard deviation to 1σ was statistically calculated based on measured data and shown in the following figures. The test facilities and engine platform are shown schematically in Figure 2. For engine performance measurements, gasoline (Research octane number, RON = 92 by CPC) was used. The fermented ethanol alcohol (by ECHO Chemical Co., Ltd.), which is guarantee reagent (GR) grade containing 5% water, was used to blend with gasoline for ethanol fuels. The blending ranged from 3% to 85% ethanol. During engine operation, the engine speed was fixed at 4500 ± 50 rpm, and the air to fuel ratio was varied to maintain stoichiometric combustion. The ignition time was considered as the crank angle for maximum pressure. The ignition time was set at the stoichiometric value for gasoline (with Research octane number, RON = 92) and was maintained during valve timing and fuel evaluations unless the engine could not be maintained at 4500 ± 50 rpm. The measurement ranges and uncertainties for major apparatus are listed in Table 2.

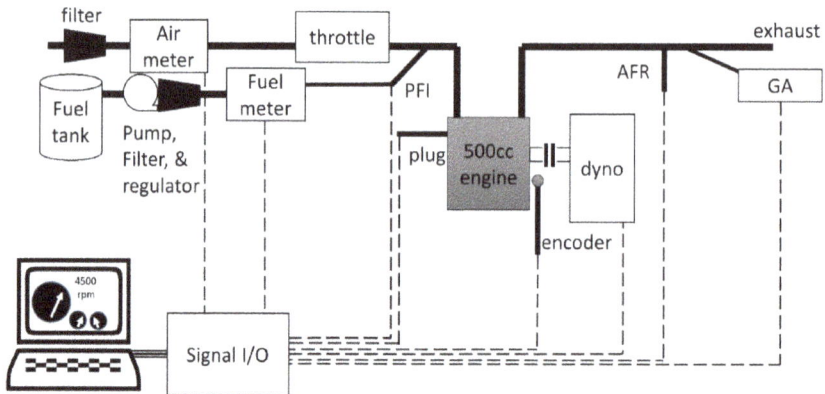

Figure 2. Single-cylinder engine mode.

Table 2. The measurement ranges and uncertainties for major apparatus.

Apparatus	Model	range	uncertainty
MEXA-584l (gas analyzer)	Co	0% to 10% vol	±0.01%
	HC	0 to 20,000 ppm	±3.3 ppm
	CO_2	0% to 20% vol	±0.17%
	NO	0 to 5000 ppm	±0.5 ppm
	O_2	0% to 25%	±0.5%
Teledyne (air mass flow meter)	HFM-200 LFE	0 to 200 LPM	±0.5%
Siemens FC (fuel mass flow meter)	MASSFLO 4100	0 to 30 kg/h	±0.1%
Optrand (pressure tranducer)	D322D6-SO	0 to 200 bar	±1%
Bosch LSU (AFR sensor)	7200	9 to 41	±2%

Engine performance was also evaluated using GT-POWER produced by Gamma Technologies Inc. [23], which is the software for simulating IC engines, capable of calculating engine performance parameters such as power, torque, airflow, volumetric efficiency, fuel consumption, turbocharger performance and matching, and pumping losses. GT-Power has been widely used by major engine manufacturers for the development of their products. A similar application of GT-Power to study the Atkinson cycle has also been proposed [24]. It was used for predicting 1-D engine performance quantities at different valve timing settings for various blends of ethanol/gasoline fuel in the present study. For 1-D engine simulation, the laminar burning velocity model of mixtures was adopted in the simulation. It has been proposed that the laminar flame speed is a function of pressure [25]. In the engine cylinder, the laminar flame speed can be estimated by Equation (2). The parameters are shown in Table 3 [26].

$$S_L(\phi, T, P) = S_{L,ref}(\phi) \left(\frac{T_r}{T_{ref}} \right)^{\alpha} \left(\frac{P}{P_{ref}} \right)^{\beta} (1 - 2.5\psi) \tag{2}$$

where

$$T_{ref} = 300K, \ p_{ref} = 1atm$$

$$S_{L,ref}(\phi) = ZW\phi^{\eta} \text{Exp} \left[-\xi(\phi - 1.075)^2 \right]$$

Table 3. Empirical coefficients for laminar flame speed.

Fuel	Z	W(cm/s)	η	ξ	α	β
C_8H_{18}	1	46.58	−0.326	4.48	1.56	−0.22
C_2H_5OH	1	46.5	0.25	6.34	1.75	$-0.17/\sqrt{\phi}$
$C_8H_{18} + C_2H_5OH$	$1 + 0.07X_E^{0.35}$	46.58	−0.326	4.48	$1.56 + 0.23X_E^{0.35}$	$X_G\beta_G + X_E\beta_E$

The single-cylinder engine model is shown in Figure 3 The general input parameters included the detailed geometry of intake and exhaust pipes, the geometries and layout characteristics of valves and cams, ignition time, fuel injection time, bore size, stroke, length of linkage, compression ratio, air to fuel ratio versus engine speed, and a user defined function for ethanol/gasoline simulation. The initial conditions for the simulation settings are listed in Table 4.

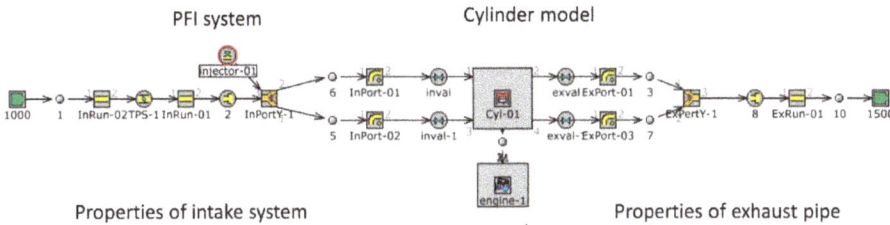

Figure 3. Single-cylinder engine model.

Table 4. Initial conditions for simulation settings.

Engine speed (rpm)	4500
Ambient pressure (Bar)	1
Ambient temperature (K)	298
Exhaust initial pressure (Bar)	1
Exhaust initial temp. (K)	1120
Exhaust initial wall temp. (K)	1000
Intake pressure (Bar)	1
Intake temp. (K)	300.15
Lamda (λ)	1.0
Intake cam lift (cm)	0.83
Exhaust cam lift (cm)	0.82

3. Results and Discussion

3.1. Setting of Valve Timing

To adjust the ECR, the timing gears were modified to meet the conditions listed in Table 5. It is noted that the top dead center (TDC) was defined as zero degree. The factory settings for the openings of the intake and exhaust valves were 314° and 86°, respectively, and this is labeled as case 0. The intake and exhaust valve opening durations were 261° and 310°, respectively. Due to the late closing of the intake valve and early opening of the exhaust valve, the effective compression and power strokes were 146° and 86°, respectively; i.e., the ECR was less than unity. In case 1, both openings of the intake and exhaust valves were delayed, but the ECR remained less than unity. In cases 2, 3, and 4, the intake opening time was maintained at 336°, and the exhaust valve closing time was tuned to 120°, 140°, and 160°. The ECR for cases 2, 3, and 4, were 0.97, 1.13 and 1.21, respectively. Note that both of the cases 3 and 4 were Atkinson cycle because their ECR was greater than unity, and others were a conventional cycle.

Table 5. Timing of valves.

Case #	Intake Valve Open	Intake Valve Close	Exhaust Valve Open	Exhaust Valve Close	Valve Overlap	Expansion/Compression
0	314	−146	86	−324	82	0.59
1	331	−128	92	−318	72	0.72
2	336	−124	120	−290	94	0.97
3	336	−124	140	−270	114	1.13
4	336	−124	150	−260	124	1.21

unit: degree

3.2. Effect of the Expansion to Compression Ratio

BSFC is an important engine parameter and allows for the comparison of different engines. It is the ratio of fuel consumption to the power produced. The real-time measured BSFC is calculated based on measured fuel consumption and work output measured by the dyno. For numerical simulation, the BSFC can be obtained in calculation results. The theoretically calculated BSFC as well as the measured results with error bars were compared and shown in Figure 4. Note that the brake specific fuel consumption is defined as the ratio of the fuel consumption rate to the power produced. According to theoretical calculations, the lowest BSFC value occurred for case 1. The measured results showed a similar trend to the theoretical calculations. For measured results, the maximum fluctuation was 7.9% for case 3. Generally, the default setting of a commercial engine does not have the best BSFC. Indeed, as the ECR was increased, the BSFC was also affectedly reduced for short stroke engines. The simultaneous heat release rate, which is defined in Equation (3), describes the relationship between crank angle and the various valve settings.

$$\dot{q} = \frac{\gamma}{\gamma-1} p \frac{dV}{dt} + \frac{1}{\gamma-1} V \frac{dp}{dt} + \frac{dQ_w}{dt} \tag{3}$$

$$Q = \int \dot{q} dt \tag{4}$$

Where $\dot{q}$ is the heat release rate (HRR), $\frac{dQ_w}{dt}$ is the convective heat-transfer rate to the combustion chamber walls, γ is the ratio of specific heats (C_p/C_v), C_v is the constant volume specific heat capacity (kJ/kg·K), C_p is the constant pressure specific heat capacity (kJ/kg·K), and Q is the total heat release (THR, kw). In addition, it has been known that γ primarily depends on the temperature and can be shown in Equation (5) [27].

$$\gamma = 1.35 - 6 \times 10^{-5} T + 1 \times 10^{-8} T^2 \tag{5}$$

The heat release rate and the total heat release rate during the single cycle of engine operation are shown in Figure 5. Higher heat release rate induces higher chamber pressure and greater force acting on the piston to produce higher power and torque. As shown in Figure 5a, the maximum heat release rate occurred in case 1. The accumulated heat release profile, which was calculated via Equation (4), is shown in Figure 5b. Despite the accumulated heat release in case 2 being similar to that in case 1, the peak in simultaneous heat release was delayed due to lower combustion flame speed as the effective expansion stroke was increased. Note that the ignition time was fixed at 315°. The power and torque corresponding to different valve timing settings are shown in Figure 6 The results show that the maximum values of both torque and power occurred in case 1 reached 21.42 N-m and 10.17 kW, respectively. The torque and power output decreased as the ECR was increased. The emissions, including CO, HC, and NO, are shown in Figure 7 The CO and HC emissions monotonically decreased as the expansion to compression ratio increased. However, the NO reached its maximum value in case 1. The intake and exhaust valve settings in case 1 showed relatively better performance, and hence were used to further study the application of bioethanol/gasoline blended fuel.

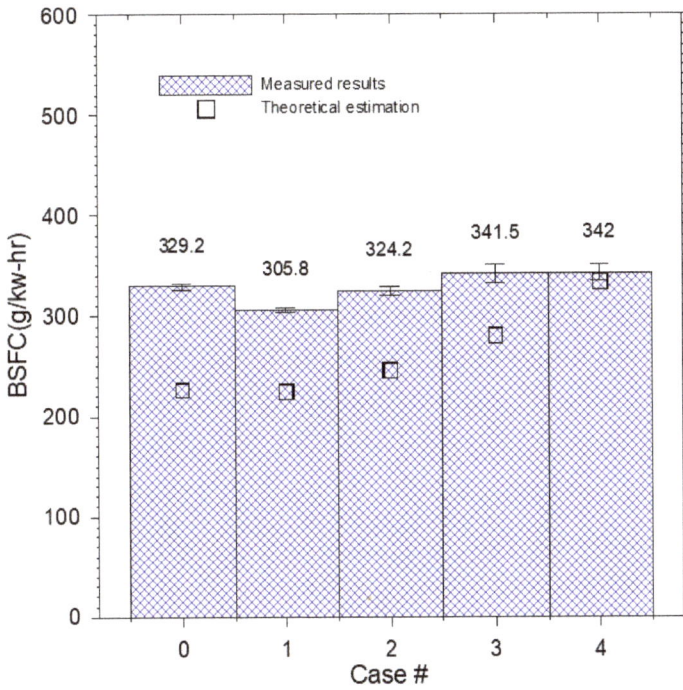

Figure 4. Comparison of measured and theoretically estimated brake specific fuel consumption (BSFC) for different valve settings.

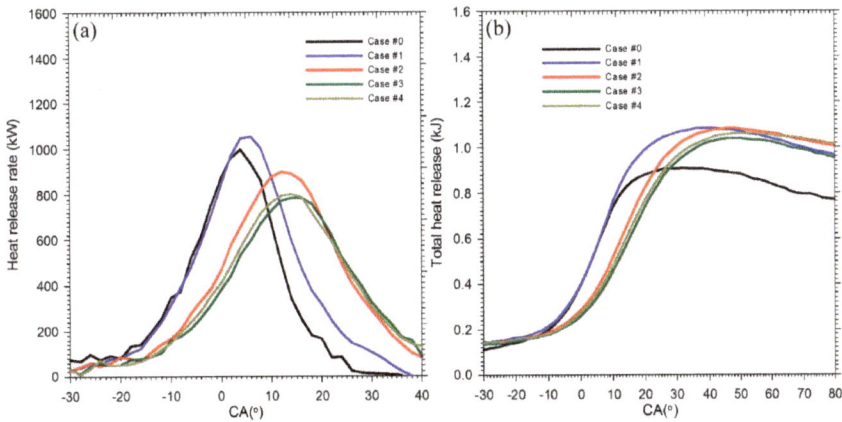

Figure 5. (a) Heat release rate; (b) Total heat release for different valve settings.

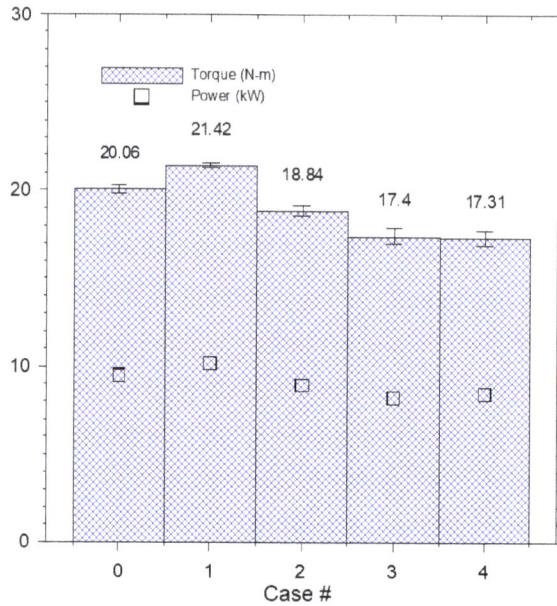

Figure 6. Power and torque output for different valve settings.

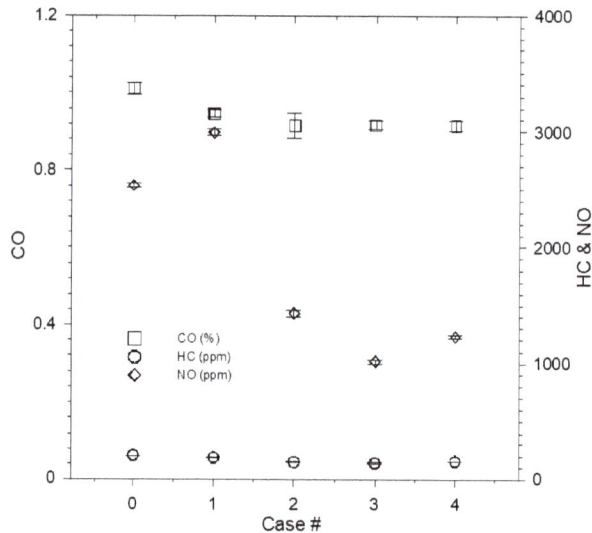

Figure 7. Emissions of CO, HC, and NO for different valve settings.

From the analysis of Figures 4–7, we can see that the engine running on case 1 conditions had the lowest BSFC value, which also indicates that the engine had the best combustion efficiency under the same fuel injection amount. This can be verified by Figure 5a for case 1. After the compression at TDC (CA 4°), the instantaneous heat release rate (HRR) reached the highest value and was significantly higher than the other cases. However, for the condition of case 1, NO generation was also the highest one, which was due to the high temperature of fuel burning in the combustion process. This result also agrees with the results presented in Figure 5a,b. Furthermore, it can be seen from Figure 4 that when the ECR was near 1.0, the internal exhaust gas recirculation value (I-EGR) of the engine can be estimated. The increase in I-EGR caused deterioration of the internal combustion efficiency of the engine, as can be seen from Figure 5, and the highest value of HRR also decreased that affected the torque output. This result also reflected in the increase of the BSFC values of cases 2, 3, and 4 when increasing the ECR values.

3.3. Bioethanol Application

Figure 8 is the corresponding BSFC trend diagram obtained by experimental measurement data and 1D numerical calculation using commercial code, GT-POWER, wherein the hollow square symbols in the figure indicate the numerical values obtained by numerical calculation. The theoretically estimated BSFC values for E3, E20, E50, and E85 demonstrate no significant change when compared to regular gasoline. The measured BSFC for E3 shows the lowest value compared to regular gasoline (E0) and other bioethenols E20, E50, and E85. It is interesting to note that with the same ignition time (315°), it was not easy to maintain steady engine operation and reach 4500 rpm using E50 and E85. Hence, the ignition time was tuned earlier by shifting to 302°. The measured results are also plotted in Figure 8 and divided by a red dashed line. As the composition of ethanol in the blended fuel was increased, the BSFC values also increased. The torque and power output for different blended fuels are shown in Figure 9. The results for E3 and E20 show an improvement in torque and power when compared to regular gasoline fuel. The superior performance of ethanol blended fuel can be attributed to the higher-octane number and the improvement in engine volumetric efficiency due to higher latent heat of ethanol. However, a higher composition of ethanol in the blended fuel induced not only a lower heating value but also lower flame propagation speed. Therefore, the performance of E50 and E85 was worse than that of regular gasoline and produced higher HC and CO emissions, as shown in Figure 10. The CO and HC emissions increased dramatically due to incomplete combustion. The various NO emissions are shown in Figure 9 and generally appeared to be lower in ethanol blended fuels compared to regular gasoline. However, E3 produced higher NO emissions than regular gasoline.

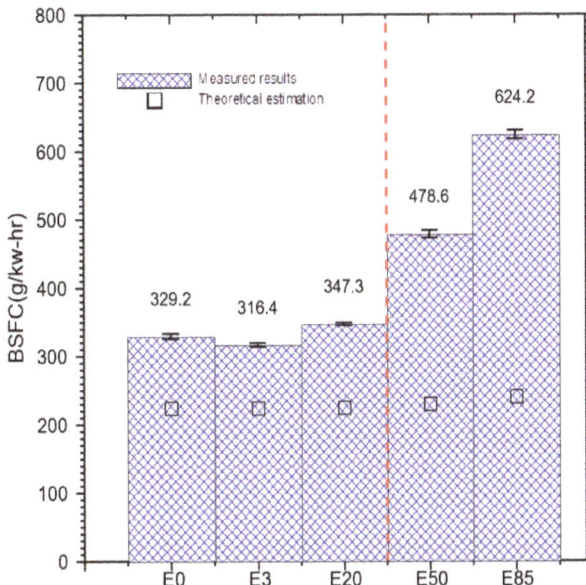

Figure 8. Estimated and measured BSFC for ethanol/gasoline blended fuel.

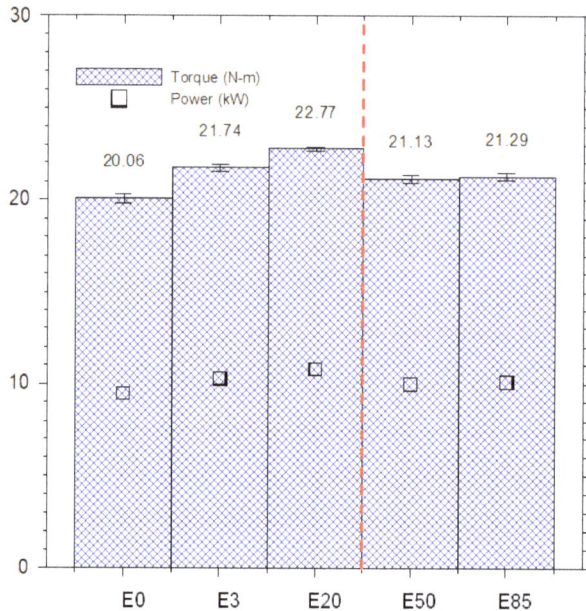

Figure 9. Power (KW) / torque(N-m) t for ethanol/gasoline blended fuel.

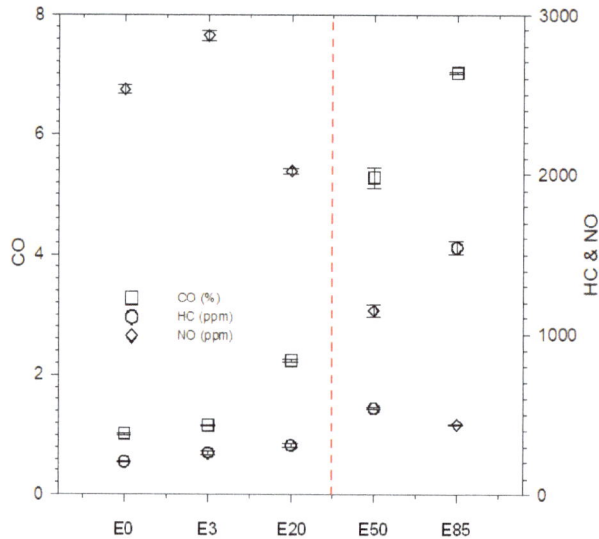

Figure 10. Emissions of CO, HC, and NO for ethanol/gasoline blended fuel.

3.4. Discussion

It has been well known that the Atkinson cycle is a method to achieve fuel economy for four stroke IC engine by controlling valve timing. Different to the vehicle applications of long stroke engine, a short stroke is more suitable for single cylinder engines and provides the advantage of higher possible engine speed with lower crank stress for motorcycle applications. In the present study, the engine speed was maintained in cruise mode and at 4500 rpm. By slightly modifying the valve timing settings, the performance and brake specific fuel consumption can be improved, at the cost, however, of increased NO emissions. The results also reveal that the brake specific fuel consumption cannot be well optimized by increasing the ECR to achieve an Atkinson cycle. In the present study, the ECR was tuned by controlling valve timing and induced a lower effective compression ratio. Due to relative shorter compression stroke as compared with the expansion stroke of the engine, more net mechanical work output can be obtained which means higher efficacy. However, a short stroke engine with a high bore/stroke ratio leads to lower thermal efficiency due to heat loss through the surfaces of the piston and cylinder heads. Therefore, thermal loss can increase as the effective expansion stroke is extended. Moreover, a fast combustion process needs to be achieved in short stroke engines. However, with a high bore/stroke ratio, the inhomogeneity of the fuel and air mixing during the combustion process results in higher emissions and lower engine performance. As the effective compression stroke is relatively short, the pressure of the fuel/air mixture prior to ignition decreases, resulting in lower flame speed and incomplete combustion, as shown in Figures 5 and 7. Due to the intrinsic properties of short stroke engines, a greater amount of fuel needs to be consumed to maintain the same engine speed if the ECR is increased further.

In addition, the heating value of ethanol was less than that of gasoline resulting in a lower flame propagation speed. The present results reveal that the ignition time needs to occur earlier to maintain steady operation in short stroke engines. In addition, CO and HC emissions increase dramatically due to incomplete combustion. Due to the intrinsic properties of ethanol/gasoline blended fuel, in order to maintain a constant engine speed with the same throttle position, more fuel is needed as the composition of ethanol in the fuel is increased.

4. Conclusions

In the present study, the effect of valve timing on engine performance was theoretically and experimentally investigated. In addition, the application of various ethanol/gasoline blended fuel (E3, E20 E50, and E85) was also studied, with a specific focus on engine performance, fuel consumption and emissions. The present study provides significant results relating to the effect of the expansion to compression ratio on the short stroke engine, and the feasibility of applying ethanol/gasoline blended fuel to a short stroke engine, i.e.,

1. With appropriate intake and exhaust valve settings, fuel consumption and engine performance can be well adjusted. However, better engine performance results in higher NO emissions. Due to the innate properties of short stroke engine, only a few operation ranges can be adjusted.
2. Short stroke engines cannot achieve high Atkinson cycle efficiency due to lower compression pressure, which induces a lower flame propagation speed and heat loss through the surfaces of the piston and cylinder head in the combustion chamber.
3. From the experimental and numerical results, we see that as the engine ECR value increases to achieve an Atkinson cycle, the engine torque, power rate, and NO emissions will all decrease, and the amount of CO and HC will increase for short stroke engines due to increased I-EGR.
4. It is necessary to vary the ignition time for fuel containing higher ethanol composition to overcome the lower flame propagation speed induced by the lower heating value of the fuel. A higher composition of ethanol in the blended fuel induces lower engine performance and higher emissions of CO and HC, while NO emissions are reduced.

The fuel consumption as well as engine performance of short stroke motorcycle engines can be improved by correctly setting the valve controls. In addition, ethanol/gasoline blended fuel can be used up to a composition of 20% without engine modification. The ignition time needs to be adjusted in fuel with higher compositions of blended ethanol. The fuel economy of a short stroke engine cannot be sharply improved using an Atkinson cycle, but CO_2 emissions can be reduced using ethanol/gasoline blended fuel. In the present study, the discrepancies between measured and calculated results can be found. Due to the limitation of 1-D simulation and the assumptions of the model, the combustion process cannot be exactly evaluated using 1-D numerical simulation. Hence, further study using 3-D numerical simulation coupled with more detailed chemical mechanism is recommended.

Author Contributions: K.-H.C. was the main author of this manuscript and contributed to the experiment design, data collection, theoretical analysis, and manuscript preparation. Y.-C.C. provided technical guidance for research work and supervised the whole project.

Acknowledgments: This research was supported by the Ministry of Science and Technology, Taiwan, R.O.C. under Grant no. MOST 105-2221-E-006 -107 -MY3 and MOST 104- 2221-E-244-008.

Conflicts of Interest: The authors declare no conflict of interest.

References

1. Statistics Inquiry, Ministry of Transportation and Communications, Taiwan. Available online: http://stat. motc.gov.tw (accessed on 20 June 2016).
2. Atkinson, J. Gas Engine. U.S. Patent 367496A.
3. Heywood, J. *Internal Combustion Engine Fundamentals*; McGraw-Hill: New York, NY, USA, 1997.
4. Ferrey, P.; Miehe, Y.; Constensou, C.; Collee, V. Potential of a variable compression ratio gasoline SI engine with very high expansion ratio and variable valve actuation. *SAE Int. J. Engines* **2014**, *7*, 468–487. [CrossRef]
5. Chen, L.; Lin, J.; Sun, F.; Wu, C. Efficiency of an Atkinson engine at maximum power density. *Energy Convers. Manag.* **1998**, *39*, 337–341. [CrossRef]
6. Wang, P.Y.; Hou, S.S. Performance analysis and comparison of an Atkinson cycle coupled to variable temperature heat reservoirs under maximum power and maximum power density conditions. *Energy Convers. Manag.* **2005**, *46*, 2637–2655. [CrossRef]
7. Hou, S.S. Comparison of performances of air standard Atkinson and Otto cycles with heat transfer considerations. *Energy Convers. Manag.* **2007**, *48*, 1683–1690. [CrossRef]
8. Boretti, A.; Scalzo, J. Novel Crankshaft Mechanism and Regenerative Braking System to Improve the Fuel Economy of Light Duty Vehicles and Passenger Cars. *SAE Int. J. Passeng. Cars—Mech. Syst.* **2012**, *5*, 1177–1193. [CrossRef]
9. Benajes, J.; Serrano, J.R.; Molina, S.; Novella, R. Potential of Atkinson cycle combined with EGR for pollutant control in a HD diesel engine. *Energy Convers. Manag.* **2009**, *50*, 174–183. [CrossRef]
10. Čuček, L.; Varbanov, P.S.; Klemeš, J.J.; Kravanja, Z. A review of footprint analysis tools for monitoring impacts on sustainability. *Energy* **2012**, *44*, 135–145. [CrossRef]
11. Mussatto, S.I.; Dragone, G.; Guimarães, P.M.; Silva, J.P.; Carneiro, L.M.; Roberto, I.C. Technological trends, global market, and challenges of bio-ethanol production. *Biotechnol. Adv.* **2010**, *28*, 817–830. [CrossRef] [PubMed]
12. Kumar, S.; Singh, N.; Prasad, R. Anhydrous ethanol: A renewable source of energy. *Renew. Sustain. Energy Rev.* **2010**, *14*, 1830–1844. [CrossRef]
13. Niven, R.K. Ethanol in gasoline: Environmental impacts and sustainability review article. *Renew. Sustain. Energy Rev.* **2005**, *9*, 535–555. [CrossRef]
14. Agarwal, A.K. Biofuels (alcohols and biodiesel) applications as fuels for internal combustion engines. *Prog. Energy Combust. Sci.* **2007**, *33*, 233–271. [CrossRef]
15. Stein, R.A.; Anderson, J.E.; Wallington, T.J. An overview of the effects of ethanol-gasoline blends on SI engine performance, fuel efficiency, and emissions. *SAE Intern. Engines* **2013**, *6*, 470–487. [CrossRef]
16. Thangavelu, S.K.; Ahmed, A.S.; Ani, F.N. Review on bioethanol as alternative fuel for spark ignition engines. *Renew. Sustain. Energy Rev.* **2016**, *56*, 820–835. [CrossRef]
17. Thompson, P.B. The agricultural ethics of biofuels: The food vs. fuel debate. *Agriculture* **2012**, *2*, 339–358. [CrossRef]
18. Chen, W.H.; Tu, Y.J.; Sheen, H.K. Impact of dilute acid pretreatment on the structure of bagasse for producing bioethanol. *Int. J. Energy Res.* **2010**, *34*, 265–274. [CrossRef]
19. Chen, W.H.; Tu, Y.J.; Sheen, H.K. Disruption of sugarcane bagasse lignocellulosic structure by means of dilute sulfuric acid with microwave-assisted heating. *Appl. Energy* **2011**, *88*, 2726–2734. [CrossRef]
20. Limayem, A.; Ricke, S.C. Lignocellulosic biomass for bioethanol production: Current perspectives, potential issues and future prospects. *Prog. Energy Combust. Sci.* **2012**, *38*, 449–467. [CrossRef]

Energies **2019**, *12*, 759

21. Kun-Balog, A.; Sztankó, K.; Józa, V. Pollutant emission of gaseous and liquid aqueous bioethanol combustion in swirl burners. *Energy Convers. Manag.* **2017**, *149*, 896–903. [CrossRef]

22. *ANSI/ASME, Part 10, Flue and Exhaust Gas Analyses*; Performance Test Code PTC 19.10; American Society of Mechanical Engineers: New York, NY, USA, 1981.

23. Gamma Technologies. Available online: https://www.gtisoft.com/ (accessed on 20 June 2016).

24. Constensou, C.; Collee, V. VCR-VVA-High Expansion Ratio, a Very Effective Way to Miller-Atkinson Cycle. In Proceedings of the SAE 2016 World Congress and Exhibition, Detroit, MI, USA, 12–14 April 2016.

25. Andrews, G.E.; Bradley, D. The burning velocity of methan-air mixtures. *Combust. Flame* **1972**, *19*, 275–288. [CrossRef]

26. Bayraktar, H. Experimental and theoretical investigation of using gasoline-ethanol blends in spark-ignition engines. *Renew. Energy* **2005**, *30*, 1733–1747. [CrossRef]

27. Brunt, M.F.J.; Platts, K.C. Calculation of Heat Release in Direct Injection Diesel Engine. *J. Engines* **1999**, *108*, 161–175. [CrossRef]

energies

MDPI

Article

Performance Analysis of Air and Oxy-Fuel Laminar Combustion in a Porous Plate Reactor

Furqan Tahir [1,*], Haider Ali [2], Ahmer A.B. Baloch [1] and Yasir Jamil [3]

[1] Division of Sustainable Development (DSD), College of Science & Engineering (CSE), Hamad Bin Khalifa University (HBKU), Education City, Doha 34110, Qatar; ahmbaloch@mail.hbku.edu.qa
[2] Department of Mechanical Engineering, DHA Suffa University, Karachi 75500, Pakistan; haider@dsu.edu.pk
[3] Department of Mechanical Engineering, King Fahd University of Petroleum & Minerals (KFUPM), Dhahran 31261, Saudi Arabia; yasirjamil@engineer.com
* Correspondence: ftahir@mail.hbku.edu.qa; Tel.: +974-30074879

Received: 12 March 2019; Accepted: 30 April 2019; Published: 6 May 2019

Abstract: Greenhouse gas emissions from the combustion of fossil fuels pose a serious threat to global warming. Mitigation measures to counter the exponential growth and harmful impact of these gases on the environment require techniques for the reduction and capturing of carbon. Oxy-fuel combustion is one such effective method, which is used for the carbon capture. In the present work, a numerical study was carried out to analyze characteristics of oxy-fuel combustion inside a porous plate reactor. The advantage of incorporating porous plates is to control local oxy-fuel ratio and to avoid hot spots inside the reactor. A modified two-steps reaction kinetics model was incorporated in the simulation for modeling of methane air-combustion and oxy-fuel combustion. Simulations were performed for different oxidizer ratios, mass flow rates, and reactor heights. Results showed that that oxy-combustion with an oxidizer ratio (OR) of 0.243 could have the same adiabatic flame temperature as that of air-combustion. It was found that not only does OR need to be changed, but also flow field or reactor dimensions should be changed to achieve similar combustion characteristics as that of air-combustion. Fifty percent higher mass flow rates or 40% reduction in reactor height may achieve comparable outlet temperature to air-combustion. It was concluded that not only does the oxidizer ratio of oxy-combustion need to be changed, but the velocity field is also required to be matched with air-combustion to attain similar outlet temperature.

Keywords: oxy-fuel combustion; porous plate reactor; oxidizer ratio; methane; CFD

1. Introduction

Fossil fuels such as furnace oil, coal, and natural gas provide 80% of the world's increasing energy demand and will continue to provide 60% of the world's energy needs by 2040 [1]. Although fossil fuel based power plants are well developed and deliver smooth operations relative to the alternative technologies such as renewables, nuclear, and carbon free systems, they have devastating drawbacks in terms of global warming, ozone depletion, and air quality [2]. Among the greenhouse gases from fossil fuel power plants, carbon dioxide (CO_2) affects the environment the most. Considering the present situation of the increasing global warming, it is very critical to limit the amount of carbon dioxide to the environment. In order to restrict CO_2 emissions, three technologies have been proposed for carbon capture and storage (CCS), namely pre-combustion, oxy-combustion, and post-combustion [3]. In pre-combustion, fuel is converted into carbon dioxide and hydrogen mixture (CO_2+H_2) and the carbon dioxide is removed from hydrogen and is sent to compression unit. Hydrogen is combusted in the presence of air for power production [4,5]. In post-combustion, flue gases from power plants are passed through selective adsorbents and absorbents to remove carbon dioxide, which is then stored or sequestrated [6,7]. The CO_2 emissions can be avoided in pre- and post-combustion technologies;

however, NO_x would still produce due to nitrogen presence in the air. In oxy-combustion, fuel is burnt in the presence of oxygen, the products include only water vapors, and carbon dioxide, considering one-step reaction, can then easily be separated by condensing the water vapors. By implementing oxy-fuel combustion technology, removal of 98% carbon dioxide from the exhaust is possible [8,9].

Oxy-combustion is an emerging technology, which can be implemented in existing plants with some modifications and new plants [10,11]. The combustion between pure oxygen and fuel results in very high temperature that the current equipment cannot withstand. Consequently, recirculation of carrier gas is required in order to reduce the maximum temperature in the equipment, and use of carbon dioxide as the carrier gas is the economic solution [12,13]. Though oxy-fuel combustion is a very promising technique for CCS, lower adiabatic flame temperature, delayed ignition, flame stability, changes in reaction kinetics, radiative heat transfer and transport properties, and lower burning rate in O_2/CO_2 environment are some of the major unresolved challenges [14–16].

Comprehensive studies have been carried out in order to evaluate potential of oxy-fuel combustion application in the existing systems [10,17,18]. Andersen et al. [19] studied chemical kinetics of methane combustion under oxy-fuel conditions. They concluded that oxy-fuel combustion characteristics are different than those of air-combustion. In their study, Westbrook and Dryer (WD) two-step, and Jones and Lindstedt (JL) four-step reaction kinetics models under O_2/CO_2 conditions were modified. Their modified mechanisms were validated against experimental data, and they found that CO prediction was improved as compared to the original models. Zhen et al. [15] performed experiments for methane combustion in N_2 and CO_2 environments, and it was observed that lift-off co-axial velocity was reduced significantly under a CO_2 environment, which indicated weaker flame stability. Zhao et al. [20] simulated laminar combustion in porous media, and their results showed better flame propagation velocities and stability as compared to an open space chamber. Kirchen et al. [10] studied ion transport membranes (ITM) for oxygen separation and oxy-fuel combustion. The ITM reactors are also termed as high temperature membrane reactors (HTMR). They concluded that oxygen permeation could be increased by using reactive gas such as methane instead of an inert gas; this would reduce oxygen separation penalties. Habib et al. [21] used ITM for oxygen separation and porous plates for CO_2 induction, in a combined ITM-porous plate reactor. It was analyzed that uniform temperature profiles could be achieved by the use of porous membranes that can enhance oxygen permeation and reactor performance. Habib et al. [22] studied oxy-fuel combustion characteristics in a porous plate reactor using methane as the fuel. They varied different parameters like oxidizer ratio (OR) and equivalence ratio (Φ) to study the impact on flame length, flow field, reaction rates, methane depletion, oxygen permeation, and maximum outlet temperature. The advantages of porous plate reactor are as follows:

1. Porous plate permeation is uniform, which helps in controlling oxidizer permeation and local oxy-fuel ratio;
2. Controlled permeation rates can avoid hotspots inside the reactor, which would allow less expensive material for reactor;
3. Permeation rate can be controlled through porosity, material, and geometrical characteristics, which would help in designing for various industrial applications;
4. Currently, permeation rates of HTMR are insufficient and porous plates can mimic HTMR conditions with higher permeation rates for future development [16].

However, for retrofitting purposes comparison with an N_2 environment in which radiative heat transfer effects, hydrodynamics, reactor geometry and transport properties are accounted for is insufficient. In the present work, performance analysis for methane combustion in a porous plate reactor under O_2/CO_2 and O_2/N_2 environments has been made. For an O_2/N_2 environment, nitrogen is used as a carrier gas with the same fraction as that of air and is termed as air-combustion. The air-combustion characteristics inside the reactor have been evaluated. For the O_2/CO_2 environment, CO_2 is used as carrier gas, and for the same adiabatic flame temperature as that of air-combustion, the oxidizer ratio has been evaluated by zero dimension analysis. For oxy-fuel combustion, the effects of oxidizer ratios, mass flow rates, and reactor height have been analyzed and compared with air-combustion (base case). In the end, parameters were highlighted that need to be modified to get the same performance as that O_2/N_2 conditions.

2. Mathematical Modeling

2.1. Adiabatic Flame Temperature

Adiabatic temperature is the maximum temperature of the flame that can be achieved theoretically under adiabatic conditions. It is important to calculate adiabatic temperature when comparing the air-combustion with oxy-combustion. In air-combustion, nitrogen is used and in oxy-combustion, CO_2 is used as a carrier gas. For this purpose, single-step stoichiometric reaction was considered for both air- and oxy-combustion that are as follows:

Air-combustion:

$$CH_4 + 2O_2 + 7.52N_2 \rightarrow CO_2 + 2H_2O + 7.52N_2 \tag{1}$$

Oxy-combustion:

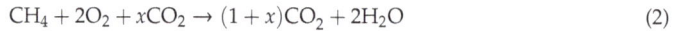

$$CH_4 + 2O_2 + xCO_2 \rightarrow (1+x)CO_2 + 2H_2O \tag{2}$$

The adiabatic flame temperature was evaluated using in-house developed code in engineering equation solver (EES) by equating enthalpies of reactants and products, then x (where x is the mole fraction of CO_2) will be calculated such that oxy-fuel reaction will have the same adiabatic flame temperature as that of air-combustion. Using the value of x, the oxidizer ratio (OR) was calculated, which represents the percentage of oxidizer in the mixture of oxidizer and carrier gas. The oxidizer ratio can be calculated by the following expression:

$$OR = \frac{\text{mass flow rate of oxidizer}}{\text{mass flow rate of oxidizer} + \text{mass flow rate of carrier gas}} \tag{3}$$

2.2. Reactor Specifications

Figure 1 represents the configuration of the porous plate reactor. Reactor has three chambers, and the middle section is the reaction zone where fuel and carrier gas enter and mixes with oxygen. Pure oxygen enters the reaction zone from the top and bottom chambers. The top and bottom chambers are closed at the end, leading all the oxygen to the reaction zone. The reaction begins when the mixture temperature is above the ignition temperature and when the mixing ratio lies within the flammability range. The height of each section is 30 mm, the width is 55 mm, and the reactor length of 500 mm. Two ceramic porous plates are used for oxygen permeation, having length of 150 mm and thicknesses of 1 mm.

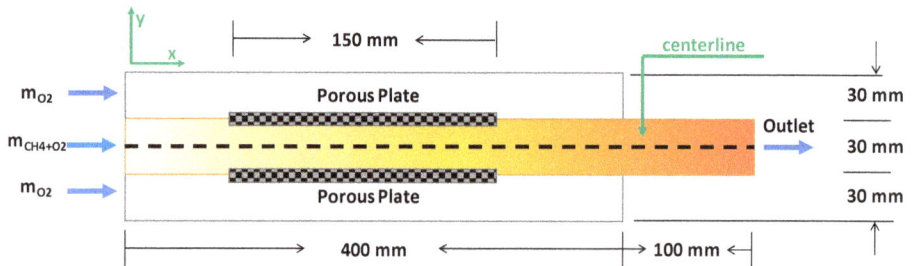

Figure 1. Schematic of porous plate reactor employed for combustion analysis.

2.3. Computational Fluid Dynamics (CFD) Model

The geometry of the reactor was constructed as a 2-D domain using commercial software, gambit 2.4. The dimensions of the porous plate were taken as 150 mm × 1mm. The x-axis is parallel to the length and the y-axis is along the height of the reactor, as shown in Figure 1. The mass, momentum, energy, and species conservation equations were used to calculate pressure (P), velocity (U), temperature (T), and species concentrations (Y). These general conservation equations are as follows:

Continuity:

$$\nabla(\rho U) = 0 \tag{4}$$

Momentum Conservation:

$$\nabla(\rho U) = -\nabla P + \mu \nabla^2 U \tag{5}$$

Energy Conservation:

$$(\rho C_p) U \nabla T = \nabla(k \nabla T) \tag{6}$$

Species Transport:

$$\nabla(\rho U T_i) - \nabla(\rho D_{i,m} \nabla Y_i) = 0 \tag{7}$$

where k is thermal conductivity and $D_{i,m}$ is the diffusion coefficient for the ith species.

Radiative heat transfer plays a vital role in all combustion processes due to elevated temperatures, i.e., around 2000 K, inside chambers or reactors. In order to incorporate the effect of radiative heat transfer, it requires the knowledge of the temperature distribution, concentration of species, and emissive properties of the medium. The radiative heat transfer lowers the limiting temperature in the reactor and stabilizes the flame. The discrete ordinates (DO) radiation model was implemented in the present study; this model can be expressed as:

$$\frac{dI(r,s)}{ds} = KI_b - (K + \sigma_s)I(r,s) \tag{8}$$

where I is the total radiation intensity, σ_s is the scattering coefficient, and K is the absorption coefficient.

For radiation modeling, the absorption coefficient of the species was evaluated using the domain-based weighted sum of the grey gas model. The internal emissivity value of all walls was taken as 0.8. The Reynolds number was kept low so that the flow was laminar inside the reactor; the laminar finite rate two-step reaction was used for the methane combustion. The reactions can be presented as follows:

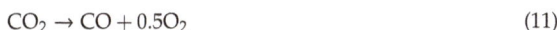

$$CH_4 + 1.5O_2 \rightarrow CO + 2H_2O \tag{9}$$

$$CO + 0.5O_2 \rightarrow CO_2 \tag{10}$$

$$CO_2 \rightarrow CO + 0.5O_2 \tag{11}$$

The rate of these reactions can be determined using the modified Arrhenius equation:

$$k_r = AT^\beta \exp\left(\frac{-E_a}{RT}\right) \tag{12}$$

where A is the pre-exponential coefficient, β is the temperature exponent, R is the molar gas constant, and E_a is the activation energy; the values of these parameters are different for air- and oxy-combustion. For air-combustion, a simple two-step reaction kinetics model was used. For the oxy-combustion study, a modified WD two-step reaction kinetics model was used. For validation, both a modified WD two-step and JL four-step kinetics models were implemented as recommended by Andersen et al. [19]. The necessary inputs for the abovementioned kinetics models are listed in Table 1.

Table 1. Values of A, β, E_a, and reaction orders for different reactions kinetics models.

Reaction	A	E_a	β	Reaction Orders
Methane-air 2 step reaction model [23]				
$CH_4 + 1.5O_2 \rightarrow CO + 2H_2O$	5.01×10^{11}	1.998×10^8	0	$[CH_4]^{0.7}[O_2]^{0.8}$
$CO + 0.5O_2 \rightarrow CO_2$	2.239×10^{12}	1.7×10^8	0	$[CO][O_2]^{0.25}[H_2O]^{0.5}$
$CO_2 \rightarrow CO + 0.5O_2$	5×10^8	1.7×10^8	0	$[CO_2]$
Westbrook and Dryer—2 step mechanism (WD) modified for O_2/CO_2 environment				
$CH_4 + 1.5O_2 \rightarrow CO + 2H_2O$	1.59×10^{13}	1.998×10^8	0	$[CH_4]^{0.7}[O_2]^{0.8}$
$CO + 0.5O_2 \rightarrow CO_2$	3.98×10^8	4.18×10^7	0	$[CO][O_2]^{0.25}[H_2O]^{0.5}$
$CO_2 \rightarrow CO + 0.5O_2$	6.16×10^{13}	3.277×10^8	−0.97	$[CO_2][H_2O]^{0.5}[O_2]^{-0.25}$
Jones and Lindstedt—4 step mechanism (JL) modified for O_2/CO_2 environment				
$CH_4 + 0.5O_2 \rightarrow CO + 2H_2$	7.82×10^{13}	1.25×10^8	0	$[CH_4]^{0.5}[O_2]^{1.25}$
$CH_4 + H_2O \rightarrow CO + 3H_2$	3×10^{11}	1.25×10^8	0	$[CH_4][H_2O]$
$H_2 + 0.5O_2 \rightarrow H_2O$	5×10^{20}	1.25×10^8	−1	$[H_2]^{0.25}[O_2]^{1.5}$
$H_2O \rightarrow H_2 + 0.5O_2$	2.93×10^{20}	4.09×10^8	−0.877	$[H_2]^{-0.75}[O_2][H_2O]$
$CO + H_2O \longleftrightarrow CO_2 + H_2$	2.75×10^{12}	8.36×10^7	0	$[CO][H_2O]$

The numerical solution is based on the CFD approach, in which a finite volume method was implemented to discretize the governing equations using commercial software fluent 18.0. In the simulation, a staggered grid was used [24]. The steady 2-D flow was simulated while using the double-precision solver. In the pressure–velocity, coupling the semi-implicit method was used. The second order upwind schemes were incorporated for the discretization of momentum and energy equation. The detailed setting regarding the simulation setup can be found in Reference [22]. Oxygen concentration and outlet temperatures were monitored to ensure the convergence. The domain was meshed with quadrilateral elements using gambit 2.4 software, a fine grid was made near walls, porous plates, and at the beginning and end of the porous plates where high gradients were expected. The grid independence study was carried out using the 16,000, 24,000, and 30,000 rectangular elements for meshing. Insignificant variations were observed for 24,000, and 30,000 rectangular elements. Therefore, for reducing computational efforts, 24,000 elements were used for the meshing of the geometry.

2.4. Operating and Boundary Conditions

The walls of the reactor are well insulated; therefore, adiabatic conditions were adopted. Mass flow inlet boundary condition and pressure outlet boundary condition were used for the inlets and outlet, respectively. Porous plates are wear-resistant and alumina-based ceramics with thermal conductivity and fluid porosity of 3.85 W/m·K and 0.55, respectively. The porous plates are considered as a 2D porous zone with viscous resistance and inertial resistance of 2.44×10^{13} m^{-2} and 100 m^{-1}, respectively. All species enter at 1173 K, i.e., higher than the self-ignition temperature of methane. Buoyancy effects have been incorporated by implementing acceleration due to gravity in negative y direction. The methane flow rate is 5×10^{-4} kg/s for the air-combustion case keeping the stoichiometric ratio. For oxy-combustion, carbon dioxide mass flow rate is varied to achieve oxidizer ratios of OR = 0.2, 0.25, 0.30, 0.35, and 0.40, while methane and oxygen flow rates were kept constant. In order to investigate the effects of mass flow rates, all flow rate values were multiplied by the factor of 1.5, 2, and 3 keeping OR = 0.25. The influence of reactor height was also examined by reducing height to $h = 24$ and 18 mm, respectively, with the flow rate conditions the same as for case 3. The details of species mass flow rates for all cases are listed in Table 2.

Table 2. Inlet conditions for the porous plate reactor (species mass flow rates).

Case #	Species Mass Flow Rate (10^{-3} kg/s)				Remarks
	CH$_4$	O$_2$/2	CO$_2$	N$_2$	
1	0.5	1	-	6.7	OR = 0.23, Air
			Effect of Oxidizer Ratio		
2	0.5	1	8	-	OR = 0.20
3	0.5	1	6	-	OR = 0.25, 1×, h = 30 mm, base case for oxy-combustion
4	0.5	1	4.67	-	OR = 0.30
5	0.5	1	3.71	-	OR = 0.35
6	0.5	1	3	-	OR=0.40
			Effect of Mass Flow Rate		
7	0.75	1.5	9	-	OR = 0.25, 1.5×
8	1	2	12	-	OR = 0.25, 2×
9	1.5	3	18	-	OR = 0.25, 3×
			Effect of Reactor Height		
10	0.5	1	6	-	OR = 0.25, h = 24 mm
11	0.5	1	6	-	OR = 0.25, h = 18 mm

2.5. Experimental Setup

The experimental setup as shown in Figure 2 was used to measure species concentration at different locations inside the porous plate reactor under non-reacting conditions. Gaseous flow rates were varied using mass flow controllers. Three types of gases were used, i.e., N$_2$ as fuel, O$_2$ as oxidizer, and CO$_2$ as carrier gas. Nitrogen (N$_2$) was selected because of non-reactive conditions, and the molecular weight of N$_2$ was comparable to CH$_4$. N$_2$ and CO$_2$ are fed from cylinders into an 800 mm long mixing chamber to ensure homogeneous mixing. The N$_2$/CO$_2$ mixture then enters the reactor as shown in Figure 1. The reactor is equipped with two porous plates of 150 mm length that start from x = 125 mm and end at x = 275 mm. O$_2$ enters from the top and bottom chambers and then passes through porous membranes and mixes with the N$_2$/CO$_2$ mixture. The samples were collected using a 5 mm diameter probe, which can be adjusted at a specific location inside the reactor by axial and transverse movement. The collected sample was then analyzed in a gas chromatograph. Further details of the experimental setup are available in [22]. The gathered data were compared with numerical results for CFD model verification.

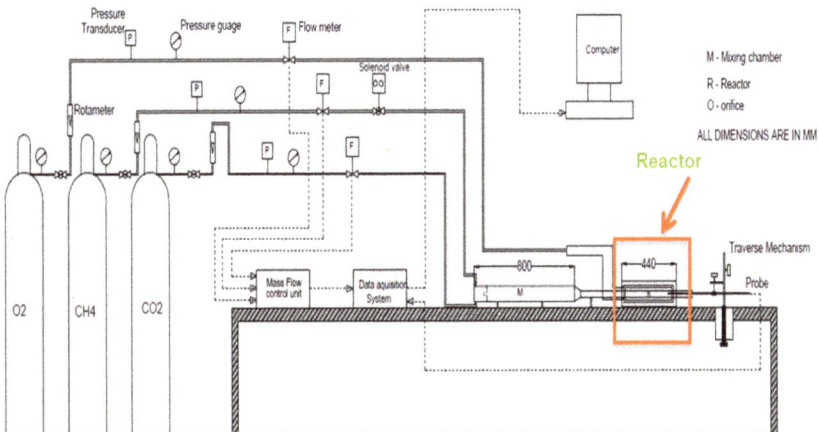

Figure 2. Schematic of the experimental setup for determining species-mixing ratios in porous plate reactor [22].

3. Results and Discussion

The CFD model for the porous plate reactor was validated in two steps, i.e., with non-reactive measurements from the porous plate reactor [22] and with reactive measurements from the gas turbine oxy-fuel combustor [25]. The local O/N ratio from the experimental data was compared with the CFD results as shown in Figure 3 and was found in good agreement. For the reactive case validation, separate geometry was made and operating conditions were set similar to Nemitallah and Habib [25]. Numerical results with a modified WD two-step and JL four-step reaction kinetics model were compared with oxy-fuel combustor data as shown in Figure 4a,b. It is evident that the four-step model better predicts the temperature profile as compared to the two-step model. The error with the two-step model is less than 12% except for one point, for which it is 20.3%. The error with the four-step model is less than 9%, except for one point, for which it is 13.5%. The deviations could be due to the accuracy of the kinetics model, turbulence model, and radiation model, 2D approximation, discretization, and uncertainties in experimental data. The deviations with the two-step mechanism are in acceptable range and thus used for numerical experiments to save computational efforts.

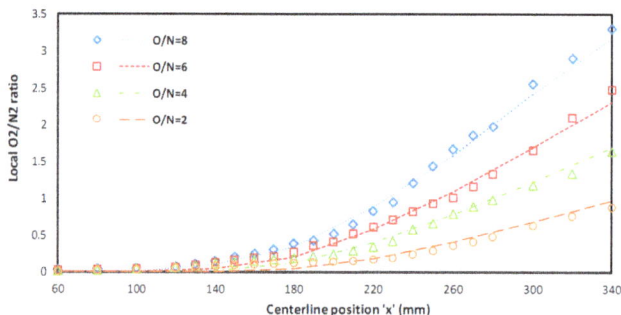

Figure 3. Comparison between experimental and numerical results for non-reacting flow inside porous reactor, at different oxygen to nitrogen ratios.

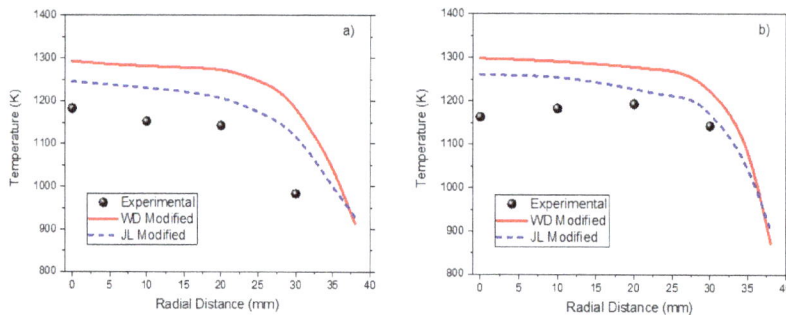

Figure 4. Comparison between the modified Westbrook and Dryer (WD) two-step, and Jones and Lindstedt (JL) numerical results with the experimental data by Nemitallah and Habib [25] under different oxy-combustion conditions in terms of exhaust temperature: (**a**) Vf = 6 L/min, O_2/CO_2 = 30/70, Φ = 0.65 and (**b**) Vf = 6 L/min, O_2/CO_2 = 50/50, Φ = 0.55.

3.1. Adiabatic Flame Temperature

The adiabatic flame temperature was found to be 2826 K for the air-combustion case that would be the maximum outlet temperature, which can be achieved in the reactor. For oxy-combustion, the oxidizer ratio OR was varied to achieve same adiabatic flame temperature as that of air-combustion and was found to be 0.243.

3.2. Air-Combustion

The average mass fraction variation of species in the reaction chamber is presented in Figure 5. The total reactor length is 500 mm, and porous plates start from $x =125$ mm and end at $x = 275$ mm. The methane mass fraction at the inlet is around 0.07, but after $x =125$ mm, methane starts to deplete as the reaction starts and diminishes at the end of porous plates. Complete methane is converted into carbon dioxide and carbon monoxide, while oxygen concentration begins to rise from zero at the start of the porous plate from where it enters the reaction chamber; the oxygen concentration rises initially and then reduces as the reaction is going on and all the oxygen is consumed. At start, water vapors mass fraction is zero and begins to increase as the reaction takes place and continues to rise until completion of reaction and water vapors mass fraction reaches a constant value. The mass fraction of carbon dioxide starts to rise as the reaction begins; it reaches a maximum value of around 0.02 and then starts to drop as some of the carbon monoxide converts to carbon dioxide. Finally, the nitrogen mass fraction has the constant value at the start but when the porous plate region starts, the oxygen induction from the porous plate into the reaction chamber reduces the overall nitrogen mass fraction.

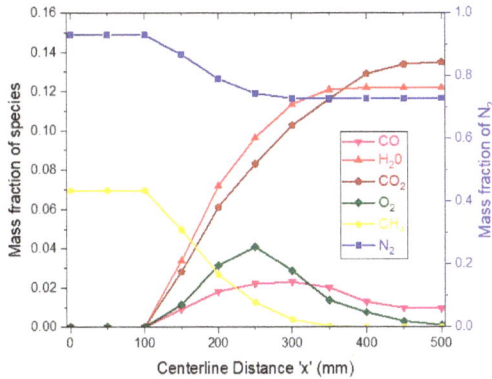

Figure 5. Variation of species mass fraction for air-combustion along the 'x' direction.

The maximum temperature in the reactor reaches 2900 K, and the outlet temperature at the exit of the reactor is 2235 K. The temperature variation inside the reactor is shown in Figure 6. The species enters the reactor at 1173 K, and the temperature of the upstream flow rises before the reaction starts, as shown by temperature level 2, i.e., 1400K; this is due to the radiative heat transfer. The region near the top and bottom porous walls has the highest temperature; this is where the combustion begins and the maximum temperature is reached. The centerline temperature is lower than the temperature near the top and bottom walls, so mass weighted average temperature variation is more accurate for study instead of centerline temperature variation.

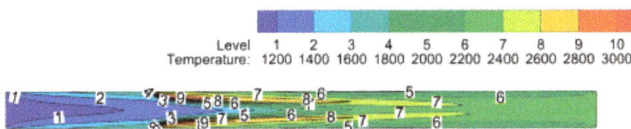

Figure 6. Temperature contours for air-combustion.

3.3. Effect of Oxidizer Ratio for Oxy-Combustion

The mass weighted average temperature variation along the 'x' axis has been plotted for oxy-combustion with OR ranging from 0.20 to 0.40, and temperature profiles are compared with the air-combustion case, as shown in Figure 7. For oxy-combustion, temperature rises sharply as the combustion occurs at the start of porous plates, i.e., x = 125 mm, where oxygen enters the reactor. The high temperature from the combustion causes upstream temperature to rise due to radiative heat transfer. As the flow temperature reaches the maximum value, it begins to decrease with an outlet temperature in the range of 2000–2050 K due to radiative cooling. The OR = 0.40 has the maximum temperature as compared to OR = 0.20, as it contains less carrier gas, i.e., carbon dioxide. Like in oxy-combustion, in the air-combustion case, temperature rises very slowly; this is because of the different transport properties of CO_2 and N_2 for convective and radiative heat transfer. Since OR = 0.243 is comparable in terms of adiabatic temperature with air-combustion and is close to OR = 0.25, the reactor outlet temperatures have a significant difference of around 200 K with air-combustion. This is due to fact that air-combustion and oxy-combustion with OR = 0.25 have comparable mass flow rates, but the density of carbon dioxide is higher than nitrogen by 1.57 times that keeps a low-velocity field in oxy-combustion and high-velocity field in air-combustion as shown in Figure 8. Inlet velocity for oxy-combustion varies from 0.32 to 0.68 m/s and inlet velocity for air-combustion is around 0.86 m/s. As the oxygen permeates through porous plates, velocity rises due to added mass flow rates and outlet velocity reaches to 2.45 m/s for air-combustion and 0.9 to 1.9 m/s for oxy-combustion. Higher velocity enhances convection heat transfer and reduces radiation effects on upstream flow, which is quite evident in Figure 7; hence, it raises the outlet temperature in the air-combustion case.

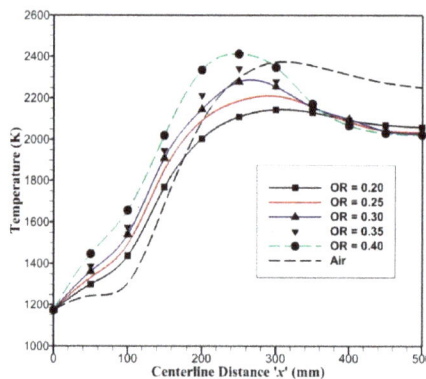

Figure 7. Mass weighted temperature variation along the 'x' direction with different oxidizer ratios for the oxy-combustion and air-combustion case.

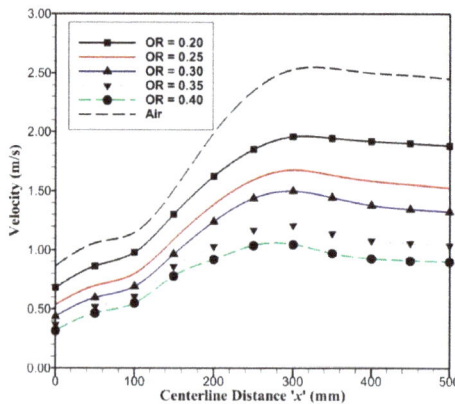

Figure 8. Mass weighted average velocity variation along the 'x' direction with different oxidizer ratios for the oxy-combustion and air-combustion case.

Figure 9a shows species mass fraction variation of methane for OR = 0.20–0.40 along the axial direction. As the OR is varied from 0.20 to 0.40, methane depletion rate increases due to a reduced velocity field in the reaction zone; which causes the residence time and temperature of species at certain point to increase, and combustion occurs faster. Due to this, complete methane depletion occurs at x = 225 mm for OR = 0.40 and at x = 325 mm for OR = 0.20. A higher methane depletion rate and less carrier gas result in higher maximum temperature in the reactor. For air-combustion, the methane depletion rate is lower due to the increased velocity field, and the reduced residence time in the reaction zone, methane diminishes completely at x = 375 mm, i.e., after porous plates x > 275 mm. Figure 9b shows species mass fraction variation of carbon monoxide along the axial direction. In air-combustion, carbon monoxide is produced only as the result of methane combustion; while in oxy-combustion, carbon monoxide is produced by methane combustion and from the disintegration of carbon dioxide at higher temperatures, which is present as a carrier gas. The carbon monoxide mass fraction at the outlet for air-combustion is lower. For oxy-combustion, higher oxidizer ratios lead to higher temperature at which carbon dioxide disintegrates into carbon monoxide and oxygen before the combustion starts, i.e., x < 125 mm. These factors result in peak shift for maximum carbon monoxide from x = 250 mm for OR = 0.20 to x = 200 mm for OR = 0.40. As the carbon monoxide reaches the maximum value, carbon monoxide starts to convert into carbon dioxide. Since for oxy-combustion cases with different ORs, the velocity field is low as compared to air-combustion, all of the carbon monoxide converts into carbo dioxide before the reactor outlet. The CO conversion rate is faster in higher oxidizer ratios. From these results, it is evident that convective effects need to be enhanced by raising the velocity field in the reactor for oxy-combustion cases to meet outlet temperature for the air-combustion case. This can be accomplished by either increasing flow rates or reducing the reactor height h.

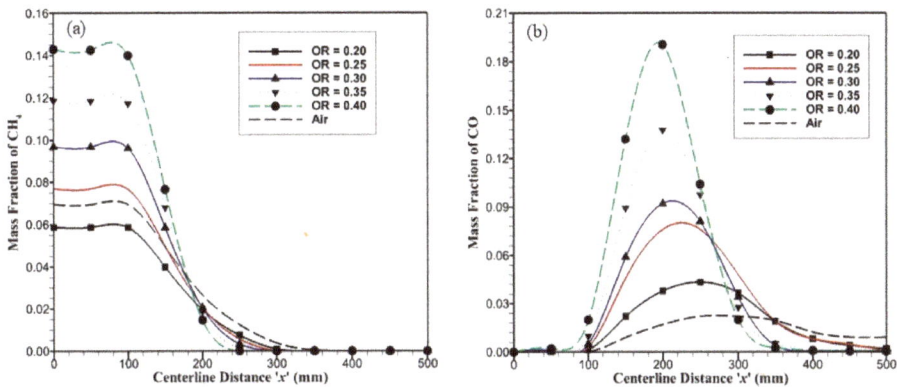

Figure 9. Variation of species mass fraction of (a) CH_4 and (b) CO, with a different oxidizer ratio for the oxy-combustion and air-combustion case along the 'x' direction.

3.4. Effect of Mass Flow Rate for Oxy-Combustion

The temperature contours at inlet flow rates (1×, 2×, 3×) are shown in Figure 10. It is noticeable from these contours that mass flow rate increment causes convective heat transfer to dominate over radiative heat transfer and upstream flow becomes cooler for higher inlet flow rates as shown by blue region. The flame length also increases as the flow rates are doubled and tripled. In addition, with higher flow rates, the flame shifts towards the reactor center due to a higher oxygen influx from porous plates.

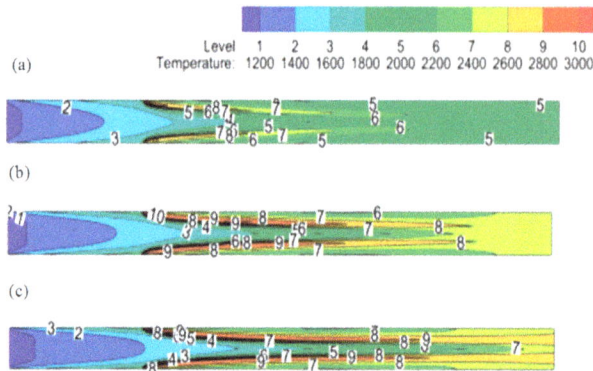

Figure 10. Temperature contours for different inlet flow rates (**a**). 1×; (**b**) 2× and (**c**) 3×.

The mass weighted average temperature variations for different inlet flow rates with fixed OR = 0.25 are shown in Figure 11a. The maximum temperature in the reactor and outlet temperature for different inlet flow conditions are shown in Figure 11b. As the mass flow rate increases, the sweep flow rate increases, enhancing the velocity field and convection heat transfer, which result in higher temperatures at the reactor outlet. The outlet temperature increases from 2034 K to around 2628 K, and the maximum temperature rises from 2855 K to 3056 K when the inlet flow rates are multiplied by a factor of 3. For 1.5× flow rates, the temperature increases to a maximum value and then slightly decreases to 2301 K, which is comparable to air-combustion with 2250 K at the outlet. However, the maximum temperature for 1.5× flow rates is lower than air-combustion due to radiative cooling, which is beneficial for reactor design. For 2× flow rates, the temperature continues to rise and maintains a constant value of around 2480 K and for 3× flow rates and the temperature continues to rise until the end and does not maintain a constant value or decrease, which reflects that the reaction is not complete by the end of the reactor.

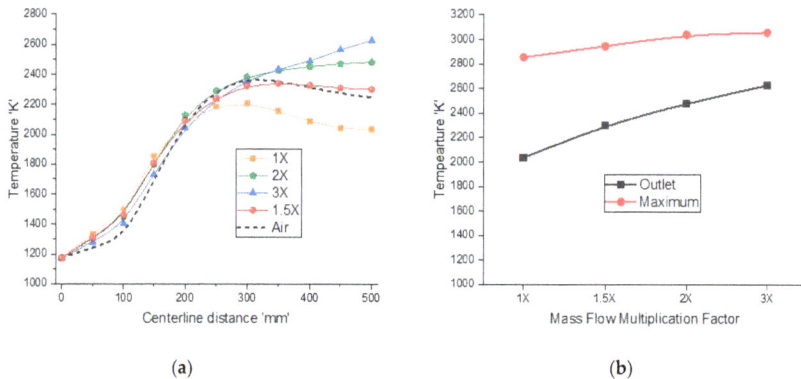

(**a**)   (**b**)

Figure 11. (**a**) Mass weighted temperature variation along the 'x' direction for different inlet flow rates with fixed OR = 0.25 (**b**) maximum and outlet temperature in the reactor for different inlet flow rates.

Figure 12 shows velocity variation along the reactor for different exit Reynolds numbers and air-combustion. As the mass flow rate is multiplied by a factor of 1.5, 2, and 3, the velocity rise is more than these factors. This is because at a higher mass flow rate, the outlet temperature increases, which lowers the average density, hence increasing the velocity field. This velocity variation behavior is similar to that of outlet temperature profiles. It is clear that the velocity fields for the air-combustion case and 1.5× inlet flow rates are similar to the temperature profiles. Therefore, for a comparable reactor performance to that of air-combustion, the mass flow rate should be increased by 50%; however, energy contents at the outlet will also be increased.

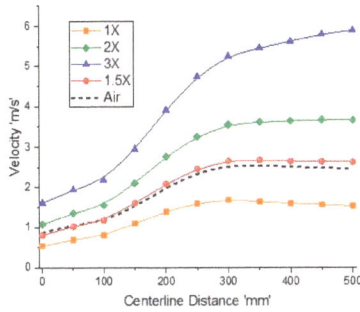

Figure 12. Mass weighted average velocity variation along the 'x' direction for different inlet flow rates with fixed OR = 0.25.

Figure 13a shows the methane depletion rate for different inlet flow rates and air-combustion. Increment in flow rates causes the velocity field inside the reactor to rise, which reduces the residence time and reaction rate. For a 1× flow rates, the reaction occurs faster, and methane diminishes at around $x = 300$ mm, while for the 1.5× flow rates, the reaction rate is slower and methane fully converts into combustible products at around $x = 320$ mm, which is comparable to that of the air-combustion case. For the 2× and 3× flow rates, the reaction is further slower, and methane converts completely at $x = 350$ mm and $x = 410$ mm, respectively. Figure 13b shows carbon monoxide fraction variation along the reaction zone for different sweep flow rates. For higher sweep flow rates, maximum CO peak shifts towards the end of the porous plate and not all of the carbon monoxide is converted and the reaction is incomplete, while for lower sweep flow rates, CO peak occurs at the start of porous plate and all of the carbon monoxide is converted before the reactor's exit. For the 3*times* flow rates, CO concentration is high as the reaction is incomplete. Since the 1.5× flow rates for oxy-combustion is comparable to that for air-combustion, they also result in similar CO mass fraction at outlet, i.e., 1% and 0.9%, respectively.

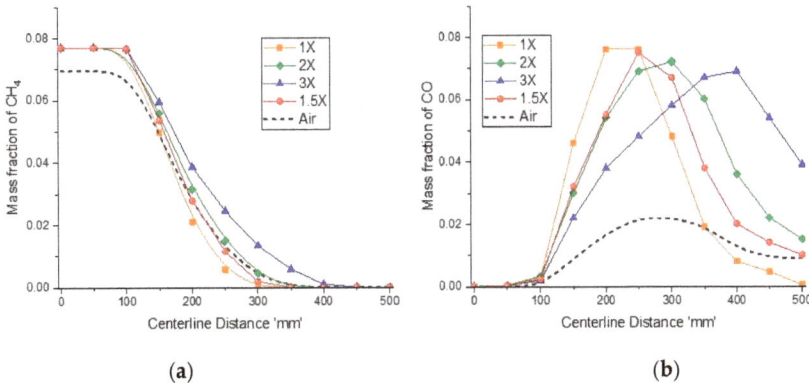

(a)

(b)

Figure 13. Variation of species mass fraction (a) CH_4 and (b) CO, different inlet flow rates with fixed OR = 0.25, along the 'x' direction.

3.5. Effect of Reactor Height

Figure 14 shows the temperature variation profiles for different reactor heights ranging from h = 30 mm to 16 mm. It can be seen that as the reactor height reduces to $h = 18$ mm, the outlet temperature changes from 2034 K to 2221K, which is a bit lower than that of the air-combustion case. However, the maximum temperature for $h = 18$ mm is much higher than that for the air-combustion. A higher temperature at a lower reactor height causes upstream temperature to rise significantly due to increased radiative heat transfer. Higher temperatures are not desirable as it limits material selection and reactor design. Figure 15a,b shows velocity and CH_4 mass fraction profile for different reactor heights respectively. Decrement in reactor heights lead to increased velocities, and a further increment in the velocity profile is observed as the combustion starts. The velocity of combustion products decreases as the mixture cools near the exit of the reactor. The higher temperature profile at $h = 18$ mm is due to the rapid combustion, as shown by the methane conversion profile in Figure 15b. The methane depletion rates for all cases are higher than methane combustion in an O_2/N_2 environment. Therefore, for a comparable reactor performance to that of air-combustion, the reactor height should be reduced by 40%; however, the peak temperature is much higher than in the air case, which is not desirable.

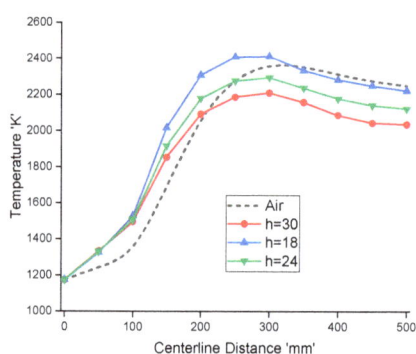

Figure 14. Mass weighted average temperature profiles for different reactor height with fixed inlet flow rates and OR = 0.25.

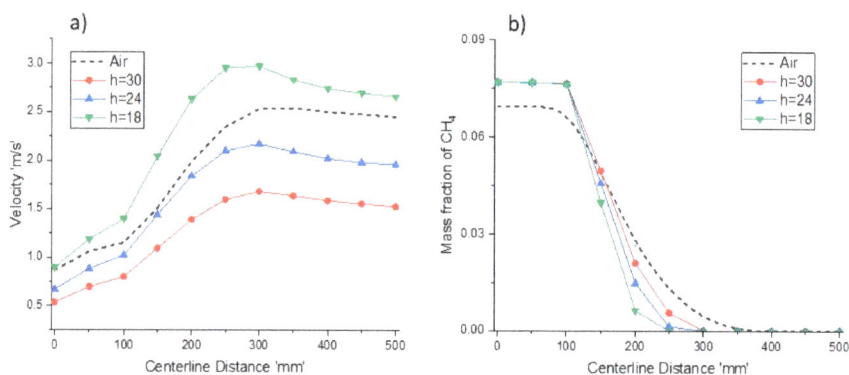

Figure 15. (a) Mass weighted average velocity profiles for different reactor height with fixed inlet flow rates and OR = 0.25 (b) CH_4 mass fraction profiles for different reactor height with fixed inlet flow rates and OR = 0.25 along the x direction.

4. Conclusions

In this study, the combustion characteristics of air-combustion were studied using a commercial CFD software fluent 18.0. It was found that the maximum temperature would be near porous plates where the reaction begins, and radiation heat transfer plays an important role in temperature distribution inside the reactor. The adiabatic flame temperature was evaluated using energy balance in engineering equation solver (EES) for air-combustion and oxy-combustion. Results showed that oxy-combustion with OR = 0.243 would have the same adiabatic flame temperature as that of air-combustion. Air-combustion was compared with oxy-combustion with different oxidizer ratios. It was concluded that the outlet temperature for different oxidizer ratios was similar and much lower as that of air-combustion because of the reduced velocity field and increased radiation effects. Oxy-combustion with OR = 0.25 was selected to study the effects of mass flow rates and reactor heights and matched with air-combustion. Higher flow rates and reduced reactor height were found to have an incremental effect on outlet temperature due to increased convective heat transfer. Furthermore, for a similar performance to that of air-combustion, the flow rate should be increased by 50%; however, total heat transfer will also be increased at the outlet. Reactor height could also be reduced by 40% to achieve similar conditions as those of air-combustion; however, the resulting higher peak temperature may not be desirable.

Author Contributions: Conceptualization, F.T. and Y.J.; methodology, F.T. and A.A.B.B.; validation, H.A.; formal analysis, A.A.B.B.; investigation, F.T. and Y.J.; writing—original draft preparation, F.T.; writing—review and editing, A.A.B.B. and Y.J.; visualization, A.A.B.B.; supervision, H.A. and Y.J.; funding acquisition, F.T.

Funding: The publication of this article was funded by the Qatar National Library (QNL).

Acknowledgments: The authors acknowledge the support provided by the Hamad Bin Khalifa University (HBKU), Qatar Foundation (QF) and King Fahd University of Petroleum and Minerals (KFUPM) to accomplish this work.

Conflicts of Interest: The authors declare no conflict of interest.

Nomenclature

A	pre-exponential factor
Cp	specific heat at constant pressure (kJ/kg·K)
CFD	computational fluid dynamics
$D_{i,m}$	diffusion coefficient of specie
DO	discrete ordinates
E_a	activation Energy (kJ/kg)
h	reactor height (mm)
HTMR	high temperature membrane reactor
I	total radiation intensity
ITM	ion transport membranes
JL	Jones and Lindstedt
$\dot{m}$	mass flow rate (kg/s)
K	absorption coefficient
k	thermal conductivity (W/m·K)
k_r	rate of reaction
OR	oxidizer ratio
P	pressure (Pa)
R	gas constant
Re	Reynolds number
T	temperature (K)
U	velocity (m/s)
WD	Westbrook and Dryer
Y	specie concentration
Φ	equivalence ratio
∇	gradient operator
β	temperature exponent
ρ	density (kg/m^3)
σ_s	scattering coefficient
μ	viscosity (Pa·s)

References

1. IEA International Energy Agency (IEA). *Key World Energy Statistics*; IEA: Paris, France, 2017.
2. Lecomte, F.; Broutin, P.; Lebas, E. *CO₂ Capture Technologies to Reduce Greenhouse Gas Emissions*; IFP Publications: Paris, France, 2010.
3. Kanniche, M.; Gros-Bonnivard, R.; Jaud, P.; Valle-Marcos, J.; Amann, J.-M.; Bouallou, C. Pre-combustion, post-combustion and oxy-combustion in thermal power plant for CO_2 capture. *Appl. Eng.* **2010**, *30*, 53–62. [CrossRef]
4. Scholes, C.A.; Smith, K.H.; Kentish, S.E.; Stevens, G.W. CO_2 capture from pre-combustion processes— Strategies for membrane gas separation. *Int. J. Greenh. Gas Control* **2010**, *4*, 739–755. [CrossRef]
5. Ali, H.; Tahir, F.; Atif, M.; AB Baloch, A. Analysis of Steam Reforming of Methane Integrated with Solar Central Receiver System. In Proceedings of the Qatar Foundation Annual Research Conference, Doha, Qatar, 19–20 March 2018; Volume 2018, p. EEPD969.
6. Merkel, T.C.; Lin, H.; Wei, X.; Baker, R. Power plant post-combustion carbon dioxide capture: An opportunity for membranes. *J. Memb. Sci.* **2010**, *359*, 126–139. [CrossRef]
7. Jamil, Y.; Habib, M.A.; Nemitallah, M.A. CFD analysis of CO_2 adsorption in different adsorbents including activated carbon, zeolite and Mg-MOF-74. *Int. J. Glob. Warm.* **2017**, *13*, 57. [CrossRef]
8. Imteyaz, B.; Habib, M.A.; Ben-Mansour, R. The characteristics of oxycombustion of liquid fuel in a typical water-tube boiler. *Energy Fuels* **2017**, *31*, 6305–6313. [CrossRef]
9. Li, M.; Tong, Y.; Thern, M.; Klingmann, J. Investigation of methane oxy-fuel combustion in a swirl-stabilised gas turbine model combustor. *Energies* **2017**, *10*, 648.
10. Kirchen, P.; Apo, D.J.; Hunt, A.; Ghoniem, A.F. A novel ion transport membrane reactor for fundamental investigations of oxygen permeation and oxy-combustion under reactive flow conditions. *Proc. Combust. Inst.* **2013**, *34*, 3463–3470. [CrossRef]
11. Chen, L.; Ghoniem, A.F. Simulation of oxy-coal combustion in a 100 kW th test facility using RANS and LES: A validation study. *Energy Fuels* **2012**, *26*, 4783–4798. [CrossRef]
12. Zdeb, J.; Howaniec, N.; Smoliński, A. Utilization of carbon dioxide in coal gasification—An experimental study. *Energies* **2019**, *12*, 140. [CrossRef]
13. Lei, K.; Ye, B.; Cao, J.; Zhang, R.; Liu, D. Combustion characteristics of single particles from bituminous coal and pine sawdust in O_2/N_2, O_2/CO_2, and O_2/H_2O atmospheres. *Energies* **2017**, *10*, 1695. [CrossRef]
14. Suda, T.; Masuko, K.; Sato, J.; Yamamoto, A.; Okazaki, K. Effect of carbon dioxide on flame propagation of pulverized coal clouds in CO_2/O_2 combustion. *Fuel* **2007**, *86*, 2008–2015. [CrossRef]
15. Zhen, H.; Wei, Z.; Chen, Z. Effect of N_2 replacement by CO_2 in coaxial-flow on the combustion and emission of a diffusion flame. *Energies* **2018**, *11*, 1032. [CrossRef]
16. Mansir, I.B.; Nemitallah, M.A.; Habib, M.A.; Khalifa, A.E. Experimental and numerical investigation of flow field and oxy-methane combustion characteristics in a low-power porous-plate reactor. *Energy* **2018**, *160*, 783–795. [CrossRef]
17. Chen, L.; Yong, S.Z.; Ghoniem, A.F. Oxy-fuel combustion of pulverized coal: Characterization, fundamentals, stabilization and CFD modeling. *Prog. Energy Combust. Sci.* **2012**, *38*, 156–214. [CrossRef]
18. Hong, J.; Chaudhry, G.; Brisson, J.G.; Field, R.; Gazzino, M.; Ghoniem, A.F. Analysis of oxy-fuel combustion power cycle utilizing a pressurized coal combustor. *Energy* **2009**, *34*, 1332–1340. [CrossRef]
19. Andersen, J.; Rasmussen, C.L.; Giselsson, T.; Glarborg, P. Global combustion mechanisms for use in CFD modeling under oxy-fuel conditions. *Energy Fuels* **2009**, *23*, 1379–1389. [CrossRef]
20. Zhao, P.; Chen, Y.; Liu, M.; Ding, M.; Zhang, G. Numerical simulation of laminar premixed combustion in a porous burner. *Front. Energy Power Eng. China* **2007**, *1*, 233–238. [CrossRef]
21. Habib, M.A.; Ahmed, P.; Ben-Mansour, R.; Badr, H.M.; Kirchen, P.; Ghoniem, A.F. Modeling of a combined ion transport and porous membrane reactor for oxy-combustion. *J. Memb. Sci.* **2013**, *446*, 230–243. [CrossRef]
22. Habib, M.A.M.A.; Tahir, F.; Nemitallah, M.A.M.A.; Ahmed, W.H.W.H.; Badr, H.M.H.M. Experimental and numerical analysis of oxy-fuel combustion in a porous plate reactor. *Int. J. Energy Res.* **2015**, *39*, 1229–1240. [CrossRef]
23. Ansys Fluent Theory Guide 12.0. Available online: http://www.afs.enea.it/project/neptunius/docs/fluent/html/th/main_pre.htm (accessed on 5 May 2019).

24. Versteeg, H.K.; Malalasekera, W. *An Introduction to Computational Fluid Dynamics: The Finite Volume Method*; Pearson Education Ltd.: London, UK, 2007; ISBN 0131274988.

25. Nemitallah, M.A.; Habib, M.A. Experimental and numerical investigations of an atmospheric diffusion oxy-combustion flame in a gas turbine model combustor. *Appl. Energy* **2013**, *111*, 401–415. [CrossRef]

![energies logo] *energies*

MDPI

Article

An Experimental and Numerical Study on Supported Ultra-Lean Methane Combustion

Ho-Chuan Lin [1], Guan-Bang Chen [2], Fang-Hsien Wu [2], Hong-Yeng Li [1] and Yei-Chin Chao [1,*]

[1] Department of Aeronautics and Astronautics, National Cheng Kung University, Tainan 701, Taiwan; edward.eas@gmail.com (H.-C.L.); calibuk88@yahoo.com.tw (H.-Y.L.)

[2] Research Center for Energy Technology and Strategy, National Cheng Kung University, Tainan 701, Taiwan; gbchen@mail.ncku.edu.tw (G.-B.C.); z10602031@email.ncku.edu.tw (F.-H.W.)

[*] Correspondence: ycchao@mail.ncku.edu.tw; Tel.: +886-6-275-7575 (ext. 63690); Fax: +886-6-238-9940

Received: 25 March 2019; Accepted: 28 May 2019; Published: 6 June 2019

Abstract: With a much larger global warming potential (GWP) and much shorter lifespan, the reduction of methane emissions offers an additional opportunity and a relatively quick way of mitigating climate change in the near future. However, the emissions from coal mining in the form of ventilation air methane (VAM), usually in ultra-lean concentration, pose the most significant technical challenge to the mitigation of methane emission. Therefore, a better understanding of ultra-lean methane combustion is essential. With three 5 mm × 50 mm rectangle cross-section slot jets, a novel sandwich-type triple-jet burner is proposed to provide stable combustion of an ultra-lean methane–air mixture with equivalence ratios from 0.3 to 0.88, and 0.22 in extreme conditions. The ultra-lean methane flame in the center of the triple-jet burner is supported by the two lean outer flames at an equivalence ratio $\varphi = 0.88$. The flow field and combustion chemical reactions are predicted by detailed numerical simulation with GRI-Mech 3.0 reaction mechanisms. Two-dimensional numerical results are validated with those obtained by experimental particle image velocimetry (PIV), as well as visual flame height and temperature measurements. An ultra-lean methane–air mixture has to burn with external support. In addition, the ultra-lean flame is non-propagating with a relatively low temperature. The ultra-lean center flame is seen to start from the outer flame and incline perfectly to the post-flame temperature and OH concentration profiles of the outer lean flame. The adjacent stronger flame provides heat and active radicals, such as OH and HO_2, from the post-flame region and in the wall proximity of the gap between the adjacent flame and the central ultra-lean jet to initiate and maintain the combustion of the central ultra-lean flame. The outstanding wall-proximity radical of HO_2 is found to be the main contributor to the initiation and stabilization of the central ultra-lean flame by providing a low-temperature oxidation of fuel through the following reaction: $HO_2 + CH_3 \Leftrightarrow OH + CH_3O$. The major chemical reaction paths contributing to fuel decomposition and oxidation of the supported ultra-lean center flame are also identified and delineated.

Keywords: mitigation; climate change; ultra-lean methane flame; lean flames; methane–air combustion; PIV; GRI-Mech 3.0

---

1. Introduction

Anthropogenic emissions of greenhouse gases (GHGs) directly altered the earth's climate and led to global warming. In addition to CO_2 as the main contributor to GHGs, the less-known non-CO_2 GHGs contributed about one-third of the total CO_2-equivalent emissions as of 2010 [1]. Among these non-CO_2 GHGs, methane contributes nearly two-thirds of present total emissions. Methane, with its abundance on earth in different forms and its low carbon ratio, is regarded as a major energy source following the fossil oil age. However, methane has a much larger global warming potential (GWP) and a much shorter lifetime (around nine years) than CO_2. Reducing methane emissions offers an additional opportunity and a relatively quick way of mitigating climate change in the near future [1]. Concentrated point sources of methane emissions from natural sources and fossil fuel exploitation are relatively important and manageable [2–4]. Most of these "fugitive methane" [4] emissions from various fossil fuel exploitations and waste treatments are released or escape without treatment, with only a smaller part being flared off. In particular, the emissions from coal mining in the form of ventilation air methane (VAM), accounting for about 6% of the total anthropogenic methane emissions, pose the most significant technical challenge to the mitigation of methane as a non-CO_2 GHG. For safety reasons, VAM usually comes in very low concentrations, usually as an ultra-lean methane–air mixture, which is defined as a mixture with a methane concentration lower than the lean flammability limit. Combustion mitigation relies heavily on high-temperature preheating, recuperative/regenerative thermal oxidation, and catalytic reactions. Practical VAM utilization in a prototype recuperative catalytic gas turbine was reported by Su and Yu [5]. Recently, combustion mitigation of methane as a non-CO_2 GHG was reviewed by Jiang et al. [4] with the emphasis on modeling and simulation of ultra-lean methane ignition/combustion. They pointed out that the challenges are faced in the ignition and sustainability of ultra-lean methane combustion and the lack of reliable chemical kinetic schemes for the development and design of advanced mitigation systems. Due to its inherent flame instability, ultra-lean methane combustion received little research attention in the past. Methods to enhance combustion stability were reported in the literature [6–10]. However, these methods deal only with lean combustion and induced instability. The application of burner arrays [11–18] and counterflow flames [19–26] for flame stability enhancement were also reported. Unfortunately, the practical ultra-lean combustion systems are complicated because of the low reaction rates, mild heat release, and low flame temperature, leading to difficulties in ignition, flame sustainability, and flame extinction. Thus, a better understanding of ultra-lean methane reaction chemical kinetics and its combustion mechanism through innovative design of the burner facility to provide stable reaction/combustion for studies is essential for this era of climate change and energy crisis.

Recently, Lin et al. [27] showed that effects of lateral impingement created by two identical adjacent slot-jet flames can lower the stable lean equivalence ratio to 0.5 or lower. The stabilization mechanism was related to the lateral impingement of post-flame streams when two adjacent slot flames were brought closer in distance. Cheng et al. [28] reported the flame structure for opposed jet flames of lean premixed propane–air versus hot products generated by lean hydrogen flames to simulate a direct injection spark ignition (DISI) engine. The lean propane–air was burnt by the support of high-temperature hot products of the lean hydrogen flame. Cheng et al. [29] further reported lean and ultra-lean stretched methane–air counterflow flames. Stretched laminar flame structures for a wide range of methane–air mixtures versus hot products of lean hydrogen flames were investigated using laser Raman diagnostics and numerical simulation. This work mostly concentrated on the comparison and evaluation of numerical codes of GRI-3.0, and C_1 and C_2 chemical kinetic mechanisms. The counterflow set-up reported in References [29,30], although good for detailed flame analysis, is much less practical for real applications. In this study, the concept of lateral flame impingement of two adjacent slot-jet flames [27] is extended to a novel sandwich-type triple-jet burner for experimental and numerical studies of the central ultra-lean flame supported through heat and radical flux provided by two adjacent lean jet flames from the hot post-flame region. Basically, the ultra-lean flame is non-self-propagating and non-self-sustainable without external support. However, the stabilization and combustion mechanism, detailed flame structure, and characteristic chemical reaction pathway associated with these supported very lean and ultra-lean flames were not comprehensively studied and, thus, represent the main purpose of this study.

2. Experimental Set-Up

2.1. Triple-Slot-Jet Burner

As shown in Figure 1a, a sandwich triple-slot-jet burner issued two identical outer jet flames and one variable center flame which were premixed and metered by two separate methane air supplies. Three equal spacing slots were used to form the sandwich triple-jet burner with a rectangular tube which had an inner cross-section of 5 mm (d, width) × 50 mm (depth). Two layers of ceramic honeycomb and one layer of steel wool were used to remove large-scale turbulent motion in the gas flow before entering the slot-jet burner. The temperature profile was measured by a coated R-type thermocouple with the junction of a 25-μm-diameter mounting on a two-dimensional (2D) moving frame driven by computer and an NI6024 PCI box in 0.02-mm increments in both x- and y-axes. This thermocouple had a deviation of 4.3% or 77.5 K at 1800 K. Two outer flames were operated identically with an equivalence ratio of 0.88. The equivalence ratio φ of the center flame varied from $\varphi = 0.7$ to 0.3. A fixed average exit velocity of 1 m/s was used for outer and center jets in all experimental cases. If the mass flow rate of the outer jets was lower, the outer flames would not produce enough heat and radicals to support the center flame and it would extinguish. In addition, according to the calculated Reynolds number (around 300), the flow field was laminar in the study. It was found that, when the centerline to centerline distance of two adjacent slot jets, L, was smaller than 2d (d = 5 mm), the triple-jet burner could stabilize the ultra-lean flame for further experiments. As a result, L was set equal to 2d for all cases in this study. The flow area near the burner jet exit was a fully developed velocity profile verified by the particle image velocimetry (PIV) measurement, as shown later in Section 4.2. There was no inert gas sheath flow around the burners. The flame appearance was captured by a digital charge-coupled device (CCD) camera, FujiPix6900. The ISO value was 400 and the aperture was F3.5.

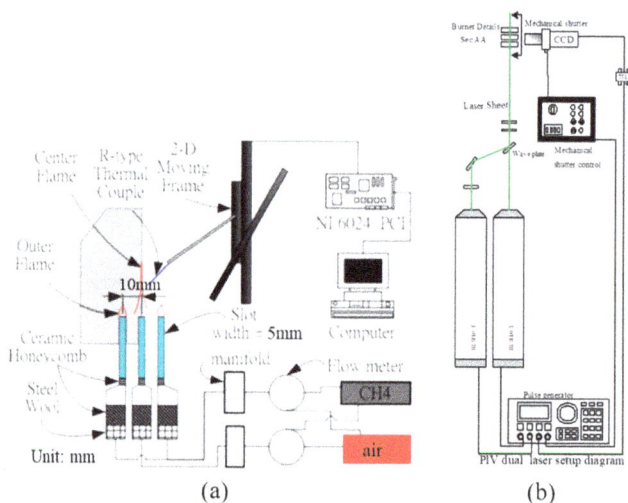

(a) (b)

Figure 1. (**a**) Schematics of the triple-jet burner and experimental measurement set-up, and (**b**) essentials of combining the shuttered particle image velocimetry (PIV) system with high-speed video camera imaging.

2.2. PIV Measurement

The flow field between the outer and center flame was carefully examined using the PIV technique. As shown in Figure 1b, the PIV system consisted of two neodymium-doped yttrium aluminum garnet (Nd:YAG) lasers, an external triggerable CCD camera, a pulse generator, a mechanical shutter, and a controller. This laser system possessed two Q-switched Nd:YAG pulse lasers (LOTIS TII LS-2134U) lasing at the fundamental wavelength (1064 nm) and second harmonic (532 nm) with a timing difference of 0.117 ms and picking pulse energy of 170 mJ. A set of laser sheet forming optics, including a wave plate, two polarizers, and cylindrical lens, was used to form a thin laser sheet of about 0.5 mm in thickness to illuminate seeding particles in the test section. A global seeding system was used with Al_2O_3 particles with a size of 7–10 µm. After mixing methane with air, the mixture flowed through a container with Al_2O_3 powder and was then introduced into the jet burner. A high-resolution and high-performance digital CCD camera (sharpVISION^{TM}) was used. Each image contained 1024×1280 pixels, which rendered a spatial resolution of 6.7 µm/pixel in the set-up. The inter-frame time for double-shutter mode could go at an interval as low as 200 ns. The mechanical shutter was used to prevent over-exposure of the second image of the PIV data image caused by strong flame illumination. The uncertainties for the major pieces of apparatus are estimated and listed in Table 1.

Table 1. The uncertainties for major pieces of apparatus. Nd:YAG—neodymium-doped yttrium aluminum garnet; CCD—charge-coupled device.

Apparatus	Model	Range	Uncertainty
Thermocouple	R-type	233–2043 K	±4.3% at 1800 K
Particle image velocimetry (PIV)	Nd:YAG pulse lasers (LOTIS TII) digital CCD camera (sharpVISION^{TM})	double-shutter mode 0–200 ns	±10%

3. Numerical Analysis

3.1. Governing Equations

To numerically model the interactive methane–air premixed flames, the governing equations of mass, momentum, energy, and chemical species for a steady reacting flow were written in the Cartesian (x, y) coordinate system as

$$\nabla \cdot (\rho \, v) = 0, \tag{1}$$

$$\nabla \cdot (\rho \, vv) = -\nabla p + \nabla \cdot (\mu \, \nabla v) + \rho \, g_x, \tag{2}$$

$$\nabla \cdot (\rho \, v \, T) = \frac{1}{c_p} \nabla \cdot (\lambda \nabla T) - \frac{1}{c_p} \sum_i h_i \{ w_i + \nabla \cdot [\rho \, D_i \nabla Y_i + \rho \, D_i^T \, \nabla (\ln T)] \}, \tag{3}$$

$$\nabla \cdot (\rho \, v \, Y_i) = \nabla \cdot [\rho \, D_i \nabla Y_i + \rho \, D_i^T \, \nabla (\ln T)] + w_i, \tag{4}$$

and the equation of state was

$$p = \rho \, R_0 \, T \sum_i \frac{Y_i}{M_i}, \tag{5}$$

where ρ, p, T, Y, c_p, h, w_i, R_0, M, g_x, and v are the density, pressure, temperature, mass fraction, specific heat capacity of the mixture, enthalpy, species production rate, universal gas constant, molecular weight, gravitational acceleration in the x-direction, and velocity vector, respectively, while μ, λ, and D are the viscosity, thermal conductivity, and mass diffusivity, respectively. The subscript i in Equations (3)–(5) stands for the i-th chemical species. The second diffusion term in the bracket of Equations (3) and (4) is the thermo-diffusion or Soret diffusion due to the effect of the temperature gradient. The concentration-driven diffusion coefficient was calculated as

$$D_i = \frac{1 - x_i}{\left[\sum_{j=1}^{N} \frac{x_j}{D_{ij}} \right]_{j \neq i}}, \tag{6}$$

where D_{ij} is the binary diffusion coefficient and the subscripts i and j stand for the i-th and j-th species, whereas the binary mass diffusivity was determined by the Chapman–Enskog kinetic theory using Lennard–Jones parameters [31]. The thermo-diffusion coefficient was calculated as

$$D_i^T = \left[\sum_{j=1}^N \frac{M_i \, M_j}{M^2} k_{ij} \, D_{ij} \right]_{j \neq i}, \tag{7}$$

where M is the mixture molecular weight and k_{ij} is the thermo-diffusion ratio.

3.2. Boundary Conditions

The relevant governing equations above were solved using the commercial package ESI-CFD for flow, heat transfer, and chemistry/mixing computations. A flame-zone reinforced grid system was used to solve the discretized equations with a control volume formulation according to the SIMPLEC algorithm. Figure 2 shows the schematic diagram of the computational domain. The x-and y-coordinates were denoted as shown in Figure 2. The origin (0,0) was located at the symmetrical centerline of the jet exit. There were four types of boundary conditions (B.C.) as indicated in the figure, i.e., the symmetrical axis, B.C. 1, B.C. 2, and B.C. 3. B.C. 1 was a fixed pressure inlet or outlet condition. For the outlet scenario, the variable gage pressure p was equal to zero and the other variables were equal to the calculated outcomes. At the inlet B.C. 2, premixed gas with a uniform velocity distribution of 1 or 0.4 m/s and a constant temperature of 300 K was ducted into the slot base. For B.C. 2, the equivalence ratio of the outer flame was fixed at $\varphi = 0.88$ and that of the center flame varied from 0.3 to 0.88. For B.C. 3, temperatures and slip conditions of the slot base and slot wall were set as constant at $T = 300$ K with non-slip conditions. The gravity force was downward as shown with g and an arrow. The domain size dimension 32 mm × 94 mm is not scaled in the figure. The heat transfer of burner slots located inside the computational domain was conjugated into the whole calculation. The appropriate boundary conditions and simulation schemes were chosen by verifying their results with the flame photo images and temperature measurements. To save computational time, the simulation began with the skeletal mechanism (16 species and 25 reactions) [32] and the results were then refined using the GRI-mech 3.0 mechanism (52 species and 325 reactions) [33]. The radiation effect was not embodied into the current numerical calculation. The assumptions and boundaries conditions used in the study were widely used for the numerical simulations of open-flame situations in the literature. We also used the same method for the simulation of methane–air combustion, and performed and discussed error analysis by comparing the results with experimental results in a similar previous study [27].

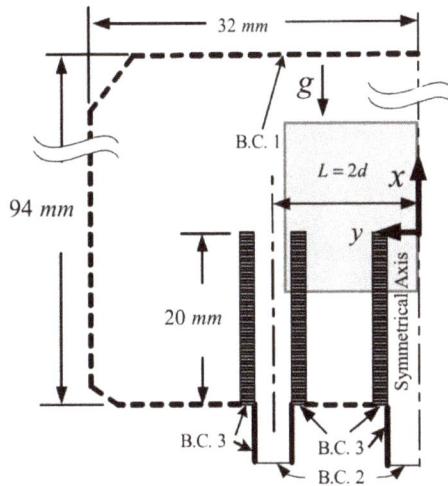

Figure 2. The computational domain and inlet and boundary conditions. The gray block near the *x*- and *y*-axes is the area shown in Figures 5–11. Boundary condition (B.C.) 1: fixed pressure (gauge $p = 0$) that can be an inlet ($T = 300$ K) or outlet; B.C. 2: uniform velocity, $T = 300$ K, premixed at $\varphi = 0.88$ for outer jets and $\varphi = 0.3$–0.88 for center jet; B.C. 3: $T = 300$ K and non-slip wall condition for all jets.

3.3. Grid and Domain Selection

This study utilized a combined grid system to generate finer grids in the flame sheet region, but kept the coarser grids for the other areas. Using different grid systems resulted in a different flame sheet contour and flame temperature. For the grid-independent study, three different grid sizes of 0.1, 0.06, and 0.03 mm within the fine grid area were tested. It was found that 0.06 mm was a reasonable grid size since the flame temperature distribution was almost the same for the grid sizes of 0.06 and 0.03 mm (as shown in Figure 3). As for the grid size of 0.1 mm, the flame temperature distribution was obviously different. The total cell number of the entire calculation zone was about 30,000. The computational time was less than one day using an HP-DL580G7 server (two Intel Xeon E7-4850 central processing units (CPUs) and 96 GB of memory).

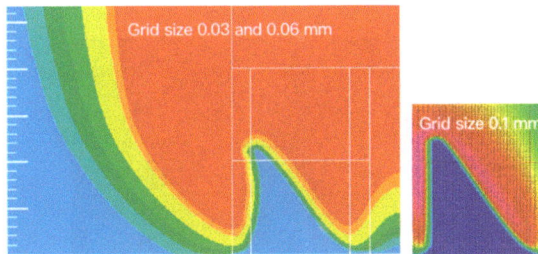

Figure 3. The temperature distribution for different grid sizes. The burning condition was at jet spacing of 2d, a burner average inlet flow speed of 1 m/s, and an equivalence ratio of 0.88.

The computational domain 32 mm × 94 mm was verified using a trial-and-error method to make sure this physical size was large enough to be a domain that was independent of size in the simulation results. A minimum required pipe length was also verified at the same time to ensure a fully developed velocity profile at the jet exit.

4. Results

4.1. Experimental Results

Figure 4 shows the flame chemiluminescence images of the sandwich triple-jet burner with the center flame varied from lean to supported ultra-lean flames. The flame chemiluminescence image was taken using a CCD camera with a filter of specific wavelength of the luminating flame intermediate species, such as OH at 307 nm. Before the camera shots, a suitable ISO value, shutter, and aperture were selected to gain the best result of the camera image. In addition, the able transform was used to analyze the brightest area and help determine the flame height. The velocity at the burner jet exit for the center and the outer premixed methane air mixtures was 1 m/s. A fixed equivalence ratio of 0.88 was applied to the outer flames for all cases, but an equivalence ratio varying from 0.3 to 0.88 was applied to the center flame. The only difference between the outer flame and center flame was the equivalence ratio. It was observed that the height of the center flame increased as the equivalence ratio decreased, especially in the cases of ultra-lean $\varphi = 0.3$ and 0.4. The tilting angle of the outer flames became less pronounced with the decrease in equivalence ratio of the center flame as shown in Figure 4a–f, especially for the ultra-lean cases ($\varphi < 0.5$). It was noted that the flame tip at $\varphi = 0.3$ was broken. The flame tip opening phenomena are related to the low Lewis number and negative stretch rate [31]. As shown in Figure 5a,b, the center slot jet was moved upstream purposely using the same burning conditions with Figure 5a for lean $\varphi = 0.88$ and Figure 5b for ultra-lean $\varphi = 0.4$ cases to distinguish the non-propagating and requiring-external-support nature of the ultra-lean flame.

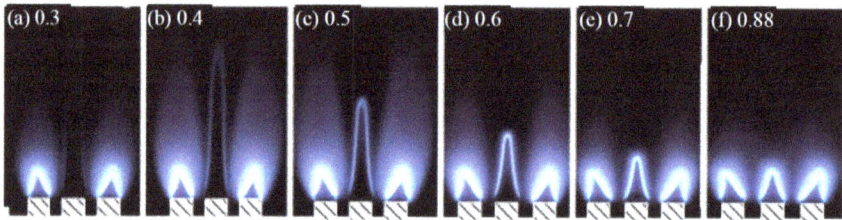

Figure 4. The flame chemiluminescence images of the outer flames for $\varphi = 0.88$ and the center flame varying from $\varphi = 0.3$ to 0.88 with equal spacing L = 2d and a constant even distribution exit velocity of 1 m/s.

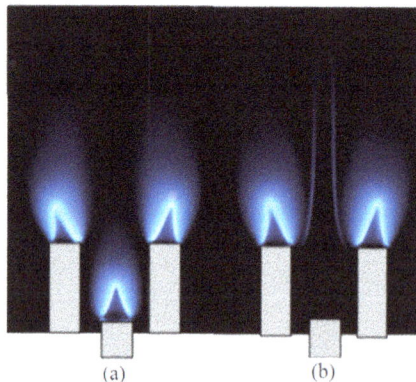

Figure 5. The flame chemiluminescence images of the triple-jet burner for the cases with the center flame equivalence ratio (**a**) $\varphi = 0.88$ and (**b**) $\varphi = 0.4$, when the center jet was intentionally placed in an upstream location.

In order to find the lowest operational equivalence ratio of the sandwich triple-jet burner, all slot spacers were removed and the remaining spacing of the jets was measured and found to be L = 1.5d due to wall finite thickness. When the jet spacing was at L = 1.5d, the operation limit of the center flame could go further down to the equivalence ratio of 0.22.

4.2. Calculated Flow Patterns and PIV Measurement

Figure 6 shows the streamline patterns in the computational domain of the enclosed gray area in Figure 2 near the jet exit, with the color-coded heat release rate distribution as the background to indicate the flame locations. The streamline pattern shows the interaction of the central and outer flames. The dividing streamline resulting from the lateral impingement of the post-flames was seen to incline to the center flame as the equivalence ratio of the center flame was reduced to ultra-lean and the flame became weak. The calculated streamline patterns of the lateral impingement showed the following characteristics: in the gap between the center and outer burner walls, the flow field was characterized by (1) the reverse flow for the case of $\varphi = 0.7$ for the center flame in Figure 6a, the reverse-and-recirculation flow for $\varphi = 0.5$ and 0.4 in Figure 6b,c, and the recirculation flow for $\varphi = 0.3$ in Figure 6d. In Figure 6, the areas in red or in green color had a heat release rate greater than 1×10^7 J·cm^{-1}·s^{-1}. The area in blue had zero heat release rate. As shown in Figure 6, upon reducing the center flame equivalence ratio, the center flame base was lifted from the burner rim, becoming close to the dividing streamline (stagnation plane of the lateral impingement) before finally going over to the other side of the dividing streamline in Figure 6c,d. However, the flame cone portion of the center flame remained on the same side for all cases. Leakage of the methane−air mixture through the lifted gap between the center flame base and the burner rim became notable for the ultra-lean cases.

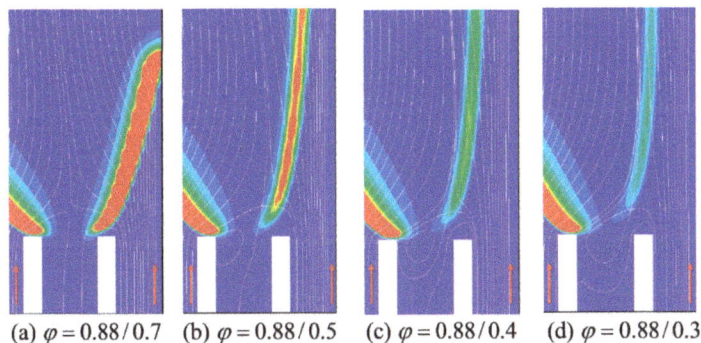

(a) $\varphi = 0.88/0.7$ (b) $\varphi = 0.88/0.5$ (c) $\varphi = 0.88/0.4$ (d) $\varphi = 0.88/0.3$

Figure 6. With the color-coded heat release rate distribution as the background, the streamline patterns show lateral impingement of the post-flame flow between the two flames with the varied center flame equivalence ratio $\varphi = 0.7$–0.3 and fixed outer flame $\varphi = 0.88$ (arrow indicates inlet flow direction).

Figure 7a shows the comparison between measurement PIV velocity vectors in yellow and the calculated streamlines in black. The burning conditions were for the center flame at $\varphi = 0.5$ and outer flame at $\varphi = 0.88$. The average jet exit speed was equal to 1 m/s for both center and outer flames. The calculated streamlines were in good correspondence with the PIV results. The stabilization mechanism of the central ultra-lean methane jet flame was highly related to the physical and chemical processes of the interaction between the central ultra-lean flame and the outer flame. Among the physical interaction processes, the velocity distribution of the combustion flow field of the interacting jet flames played an important role in the stabilization of the ultra-lean flame, and the good agreement between the CFD velocity results and PIV results served to partially validate the CFD predictions. Figure 7b compares the calculated velocity with the experimental PIV data along a section perpendicular to the center ultra-lean flame from the symmetrical axis at the position of 5 mm above the slot exit. The velocity trends matched well with each other, and the maximum deviation was about 18%. Currently, it is difficult to avoid the mutual influence and interference of these flames when experimentally measuring the species information. Some useful experimental quantities of laser combustion diagnostics of the concentration and radical (such as OH) information may be considered for further detailed studies in the future.

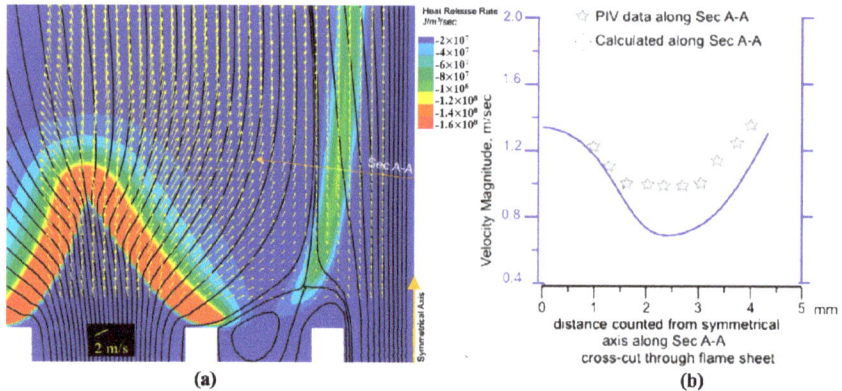

Figure 7. (**a**) Comparison between PIV velocity vectors in yellow and the calculated streamlines in black with the color-coded heat release rate distribution as the background. The burning conditions were for the center flame at $\varphi = 0.5$ and outer flame at $\varphi = 0.88$. The average jet exit speed was equal to 1 m/s for both center and outer flames. (**b**) Comparison between PIV data and calculated velocity.

Similar to Figure 6, the background of Figure 7a shows the calculated heat release rates to indicate the flame position. It was noted the deflection angles of streamlines across the outer flame and the center flame were quite different, characterizing a different flame temperature and strength.

It was reported in Reference [27] that the lateral impingement of the post-flame streamline is useful to stabilize the flame when two adjacent slot-jet flames are brought close to each other. Thus, only the typical condition of stable flames with the jet spacing L = 2d is discussed in this work.

4.3. OH Concentration

Figure 8 depicts the calculated OH concentration in white isopleths against the background of heat release rate, which was used to represent flame location. The white rectangular blocks represent the slot-jet wall of the outer and center jets. The right side of the picture is lined up with the symmetrical line and x-axis. The black streamlines are shown in Figure 8 to indicate the position of diving streamline and flow pattern. As shown, the center flame was perfectly enveloped within two OH concentration isopleths 1 (1×10^{-5} mol/m^3) and 2 (3.1×10^{-4} mol/m^3) rooted from the outer flame. If being tracked from the base of the outer flame, the distance between OH isopleth 1 and 2 got wider and then narrower after reaching the center flame base. In contrast, the distance between isopleth 2 and 3 constantly got wider. In other words, the OH diffusion strength across OH isopleths 1 and 2 became weaker from the outer lean flame base over the impingement area and became stronger along the center flame sheet, which could be identified by tracing the distance between these two isopleths, as shown in Figure 8. This indicates that the base of the center flame received OH radicals from the outer flame. However, the main portion of the center flame did locally generate a small amount of OH radicals via its own reactions. This is one of the reasons why the center flame was attached to the outer flame and not to the center jet burner rim when the center slot jet was moved upstream in Figure 5. In other words, the existence of the ultra-lean center flame depended on the support of OH radicals from the post-flame of the outer flame.

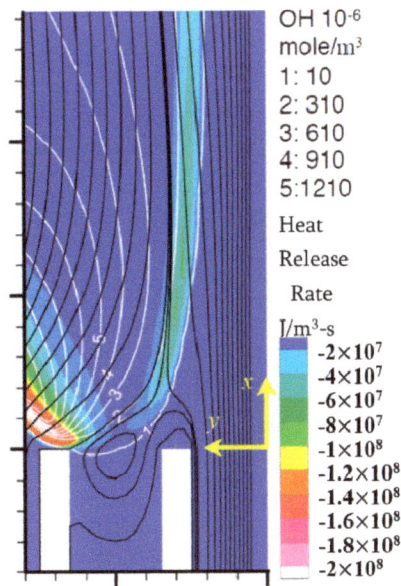

Figure 8. OH isopleths in white and streamlines in black in the post-flame region in contrast to the color-coded heat release rate distribution. The burning conditions were for the center flame at $\varphi = 0.5$ and outer flame at $\varphi = 0.88$. The average jet exit speed was equal to 1 m/s for both center and outer flames.

4.4. OH and CO Diffusion Vectors

Figures 9 and 10 show the OH and CO diffusion vectors. The symmetry axis is along the *x*-axis. The white lines indicate the streamlines. In the background of these two figures, the heat release rates of the center flame in green and outer flame in both red and green are drawn to indicate the position of the outer and center flames, and the blue color means zero heat release rate in that area. The white OH diffusion vectors pointed away from the red area or higher-concentration area in the outer flame, but there was no change in direction observed for white OH diffusion vectors over the center flame. Note that the OH diffusion vectors all went toward the symmetry axis regardless of which side they were on with respect to the center flame. In other words, the reaction of the center flame depended on the diffusion of the OH radicals from the post-flame of the outer flame. However, the CO diffusion vectors went against each other from the outer and center flames, as shown in the middle of Figure 10, indicating that CO was generated locally from both the center and outer flames.

0.1 g/m²s OH

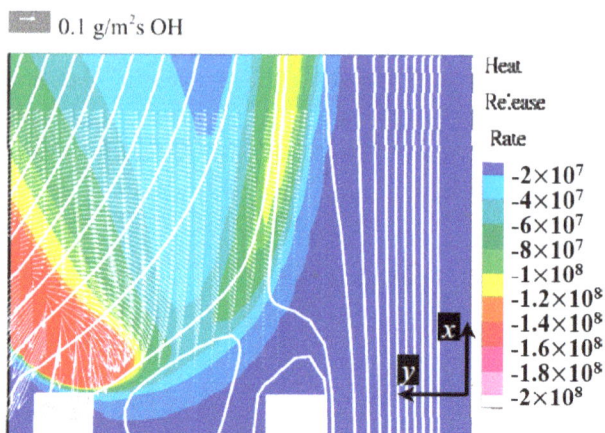

Figure 9. The OH diffusion vector (arrow) distribution in the post-flame region between the outer and the center flame, in contrast to the streamline (line) pattern and color-coded heat release rate. The burning conditions were for the center flame at φ = 0.5 and outer flame at φ = 0.88. The average jet exit speed was equal to 1 m/s for both center and outer flames.

0.5 g/m²s CO

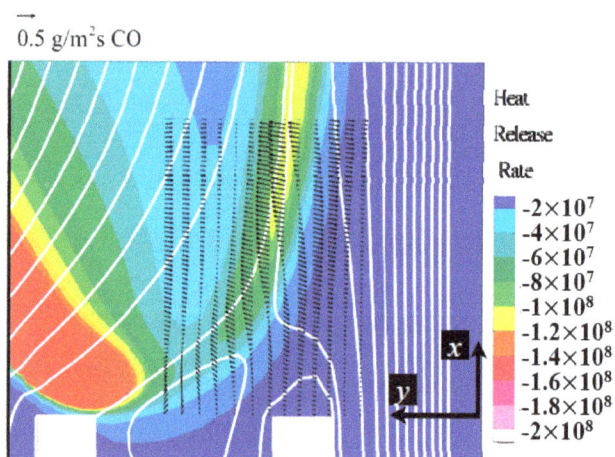

Figure 10. The CO diffusion vector distribution in the post-flame region between the outer and the center flame, in contrast to the streamline (line) pattern and color-coded heat release rate. The burning conditions were for the center flame at φ = 0.5 and outer flame at φ = 0.88. The average jet exit speed was equal to 1 m/s for both center and outer flames.

4.5. Flame Shape

Figure 11 shows the side-by-side comparison of flame location, flame length, and flame shapes between the simulated image of heat release rate and the photo picture of the triple-slot-jet flame. The burning condition of this case was an equivalence ratio of 0.88 for both outboard flames and 0.5 for the center flame, and a burner average exit speed of 1 m/s for all three of the flames. In order to acquire the best result of the camera image of the center flame, the camera was aimed at the mid-point of the center flame. At this position, the camera was aimed at the top of the outboard flames. Thus, the image of outboard flames looks slightly fuzzy or slightly out of focus. However, in this way of image shooting for the center flame, it was shot in a near-perpendicular position so that the photo could be directly compared with the simulated result in a side-by-side position without distortion. It was found that the simulated result matched well with the photo image. The measured flame height of the center ultra-lean flame was about 26 mm, and the calculated flame height was about 26.2 mm. The deviation was only about 0.77%.

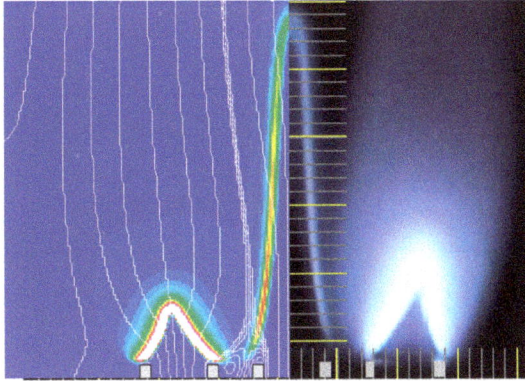

Figure 11. The triple-jet burner is illustrated; the right half shows the picture and the left half shows the simulated image at a burner average exit speed of 1 m/s and an equivalence ratio of 0.88/0.5/0.88 for the outboard/center/outboard.

4.6. Calculated and Measured Temperature Isopleths

The calculated temperature isopleths for the center flame $\varphi = 0.4$ case are compared with thermocouple measurements and a flame image in Figure 12. The comparison with flame image is shown in Figure 12a, with the thermocouple measurement in Figure 12b. The calculated temperature isopleths (lines 1 (400 K), 2 (800 K), 3 (1200 K), and 4 (1600 K)) were superimposed on the flame image and experimental temperature profile. The calculated temperature profile was in good correspondence with the measured results and flame image. The measured maximum flame temperature was 1853 K while the calculated value was 2005 K. The deviation was about 8%, which was probably due to the uncertainty of the R-type thermocouple (±4.3% at 1800 K) and heat radiation to the environment. The temperature profile of the ultra-lean center flame of $\varphi = 0.4$ lay between temperature isopleths 2 (800 K) and 3 (1200 K) as shown in Figure 12a,b, denoting a relatively low-temperature "flame". The appearance of the center flame in the flame photograph also had very good correspondence with the temperature isopleths in Figure 12b, indicating the close relationship between the ultra-lean center flame and the hot product gas of the outer flame. In view of the relatively low-temperature and non-propagating nature of the ultra-lean flame, it is argued that this "flame" is not really a flame; however, with external support chemical reactions of species conversion (e.g., CO generation), although relatively small, heat release does exist in a thin layer. Further discussions on the flame structure and major chemical reaction path are presented in Section 5.

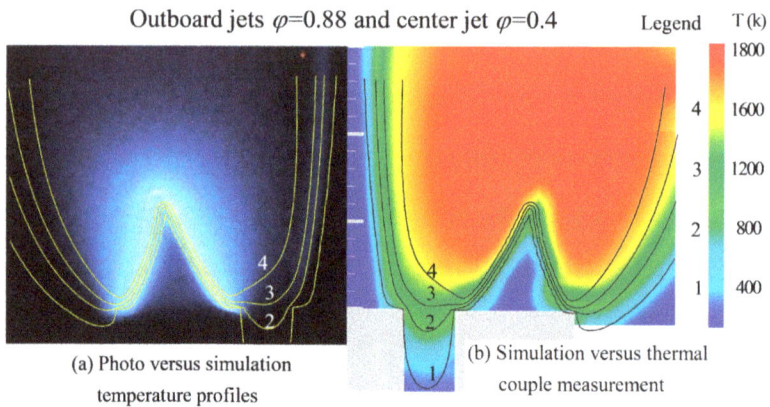

Outboard jets φ=0.88 and center jet φ=0.4

(a) Photo versus simulation temperature profiles

(b) Simulation versus thermal couple measurement

Figure 12. Comparisons of the calculated temperature isopleths against (**a**) the photo and (**b**) the experimental thermocouple measured temperature distribution of the outer flame and half of the center flame. The burning conditions were for the center flame at φ = 0.4 and outer flame at φ = 0.88. The average jet exit speed was equal to 1 m/s for both center and outer flames.

5. Discussions

Although the above simulation results showed good agreements when compared with the experimental data, one should estimate the computational error and carefully examine the assumptions and boundary conditions used in the numerical simulation before further discussions on the supported ultra-lean flame stabilization and chemical reaction characteristics using simulation results. For ease of comparison, the simulation boundary conditions and assumptions, and the resultant maximum deviation as compared with experimental measurements are summarized and listed in Table 2. The discrepancy of the predicted flame appearance, especially the flame height of the ultra-lean flame, was about 1%, and the maximum deviation in temperature prediction was about 8% with an uncertainty of ±4.3% for the R-type thermocouple at high temperature. Therefore, the constant 300 K inlet temperature boundary condition and the assumption of neglecting thermal radiation used in the numerical simulation are justifiable since the current ultra-lean flame and the supporting lean flame (at φ = 0.88) were light-blue in color and non-sooting (see Figure 4). The velocity discrepancy was relatively large with a maximum error of 18% relative to the PIV measurement. However, the uncertainty of the PIV measurement was estimated to be about 10%. From Figure 5 and the results above, as well as further discussion on the ultra-lean flame stabilization mechanism below, we show that the ultra-lean flame is too weak and the stabilization mechanism is mostly dominated by the chemical reaction supported by the supporting flame from the post-flame. Therefore, the flame shape and the temperature are the key factors related to the ultra-lean flame stabilization mechanism, and velocity discrepancy may not have a strong influence on the stabilization.

Table 2. The error analysis for simulation boundary conditions and assumptions.

Boundary Conditions (BC) and Assumptions	Parameters	Maximum Deviation to the Measured Data
• Premixed gas with uniform velocity for inlet BC	Flame height (mm) *	1% (Figure 11)
• A constant temperature of 300 K for inlet BC		
• A fixed gage static pressure $p = 0$ for inlet BC	Temperature (K)	8% (Figure 12)
• A fixed gage static pressure $p = 0$ for outlet BC		
• Thermal radiation is neglected	Velocity (m/s)	18% (Figure 7)

* Flame height was determined by the photo picture (experiment) and image of heat release rate (simulation).

One of the main objectives of this study was to delineate the stabilization and combustion mechanism of the ultra-lean methane jet flame supported by two adjacent lean jet flames. The stabilization mechanism of the central ultra-lean methane jet flame was highly related to the physical and chemical processes of the interaction between the central ultra-lean flame and the outer flames. However, the central ultra-lean jet flame was relatively too weak, and the existence and stabilization of the central ultra-lean flame depended significantly on the thermal and radical flux from the adjacent stronger flame, as can be seen from the fact that the central ultra-lean flame was almost attached to the high-temperature post-flame of the adjacent jet flame in Figures 4 and 5. Therefore, the contribution to the ultra-lean flame stabilization from the physical process in terms of flow velocity interaction between two adjacent flames was relatively low compared with the chemical process. In other words, the stabilization of the ultra-lean central flame depended more on the chemical process of the interaction by providing heat and active radicals from the post-flame region of the adjacent flame to initiate and maintain the combustion of the central ultra-lean flame. Therefore, there was more of a chemical/combustion mechanism than physical mechanism in the stabilization. In order to further delineate these cross-stream chemical activities, chemical sensitivity analysis for the reaction rates [31] was performed based on the results of four major reactions; R99: $OH + CO \Leftrightarrow H + CO_2$, R86: $OH + OH \Leftrightarrow O + H_2O$, R38: $O_2 + H \Leftrightarrow O + OH$, and R98: $OH + CH_4 \Leftrightarrow CH_3 + H_2O$ were identified as the key reaction steps for the ultra-lean combustion. Most important of all, among these four reactions, the OH radical was involved in the forward and/or backward reactions, indicating that the OH radical is the key to ultra-lean flame reactions. OH is involved in the initial fuel decomposition or hydrogen abstraction process from methane, as indicated in R98. Methane can be decomposed initially by reactions R98: $OH + CH_4 \Leftrightarrow CH_3 + H_2O$, R53: $H + CH_4 \Leftrightarrow H_2 + CH_3$, and R11: $O + CH_4 \Leftrightarrow OH + CH_3$. Upon decreasing the equivalence ratio, the weights of these three reactions shift. For the center flame, the weights of relative contribution to CH_4 decomposition initiation reactions were about 75%, 7%, and 18% for R98, R53, and R11, respectively. OH is also involved in further heat release of CO oxidation through R99. Reactions 86 and 38 show the branching reactions, in which R86 indicates that OH may be involved in providing the active oxidation radical of O atom through a different pathway other than high-temperature decomposition of oxygen molecules. As shown in Figures 8, 9 and 12, highlighting the temperature distribution, the OH concentration distribution, and the OH diffusion vectors, it can be seen that the ultra-lean flame was relatively weak with a temperature around 1200–1400 K and OH was mainly dependent on diffusion from the post-flame region of the adjacent stronger $\varphi = 0.88$ flame.

To further look into the chemical reactions related to the process of how the adjacent stronger flame provides active radicals to stabilize the ultra-lean flame in the flame base region, reaction sensitivity analysis was further performed and confined to the wall-proximity region of the gap between the adjacent flame and the central ultra-lean jet. The process in the flame base region is strongly related to initial oxidation of the fuel and maintaining the combustion, leading to stabilization of the weak central ultra-lean flame. Among the weak reactions, R119: $HO_2 + CH_3 \Leftrightarrow OH + CH_3O$ was identified as the key reaction in this region. Figure 13 shows the reaction rate distribution of R119. The outstanding wall-proximity radical of HO_2 is seen (in Figure 13) to distribute from the post-flame of the adjacent outer jet flame to directly connect the flame base (flame stabilization location) of the central ultra-lean flame, leading to the initiation and stabilization of the central ultra-lean flame by providing a low-temperature oxidation of fuel through reaction R119: $HO_2 + CH_3 \Leftrightarrow OH + CH_3O$, rather than the high-temperature oxidation through decomposition of the oxygen molecule as seen for conventional methane combustion. Figure 13b shows the enlarged plot of the jet exit wall-proximity region with a schematic expression of the reaction process of the ultra-lean flame near the flame base.

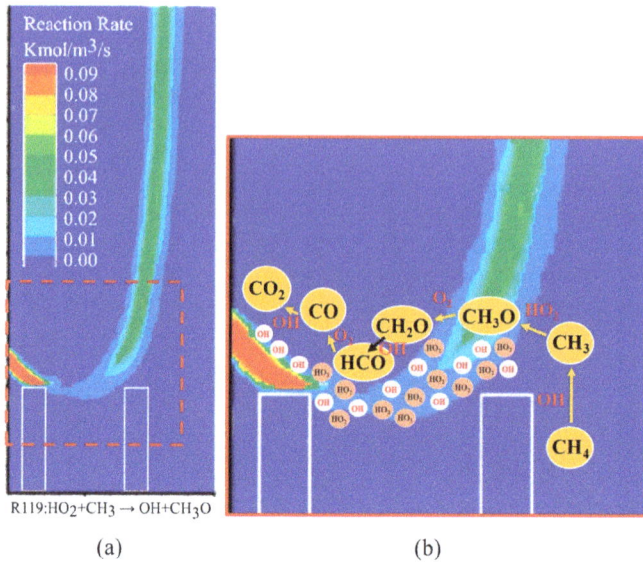

$R119{:}HO_2{+}CH_3 \rightarrow OH{+}CH_3O$

(a)   (b)

Figure 13. (**a**) The reaction rate distribution of the outstanding reaction of R119: HO_2 + CH_3 ⇔ OH + CH_3O, and (**b**) the enlarged plot of the jet exit wall-proximity region with a schematic expression of the reaction process of the ultra-lean flame near the flame base.

The above arguments and discussions can be summarized with a comparison of the major reaction pathway of the supported ultra-lean methane jet flame with respect to the conventional stable lean and stoichiometric methane flame in Figure 14. In Figure 14, in reaction R98 OH reacts with CH_4 to produce CH_3 and H_2O. For the lean or stoichiometric methane flame, the H and O are also important for the decomposition of CH_4. Instead of O, HO_2 reacts with CH_3 to CH_3O in the ultra-lean flame through R119: HO_2 + CH_3 ⇔ OH + CH_3O, and, for the lean or stoichiometric methane flame, CH_3 may decompose directly to CH_2O through a reaction with the O atom. For the ultra-lean center flame, CH_3O is further attacked by the third body M to form CH_2O, and CH_2O then reacts with OH again to produce HCO. Reaction R168: HCO + O_2 ⇔ HO_2 + CO takes HCO and O_2 together to produce CO. In these two steps, from CH_2O to HCO and then further down to CO, the center flame has fewer choices than the outer flame. After that, CO reacts with OH to form H and CO_2, which is the same for both center and outer flames. With the relative abundance of the OH radicals, with respect to the H and O radicals in the post-flame of the outer flame, as shown by the relative weights above, the initiation, decomposition, and subsequent reactions of methane in the ultra-lean center flame are mainly dominated by the OH radical and its participating reactions. With the lower flame temperature, the ultra-lean center flame tends to take the slower HO_2 route for further CH_3 reaction down to CH_3O. However, the main route of methane oxidation remains the same for both the center and the outer flames.

Sub-limit lean	Lean or stoichiometric
CH_4	CH_4
$\downarrow$ + OH	$\downarrow$ + OH, H, O
CH_3	CH_3
$\downarrow$ + HO_2	
CH_3O	$\downarrow$ + O
$\downarrow$ + M, O_2	
CH_2O	CH_2O
$\downarrow$ + OH	$\downarrow$ + OH, H, O
HCO	HCO
$\downarrow$ + O_2	$\downarrow$ + O_2, M, H
CO	CO
$\downarrow$ + OH	$\downarrow$ + OH
CO_2	CO_2

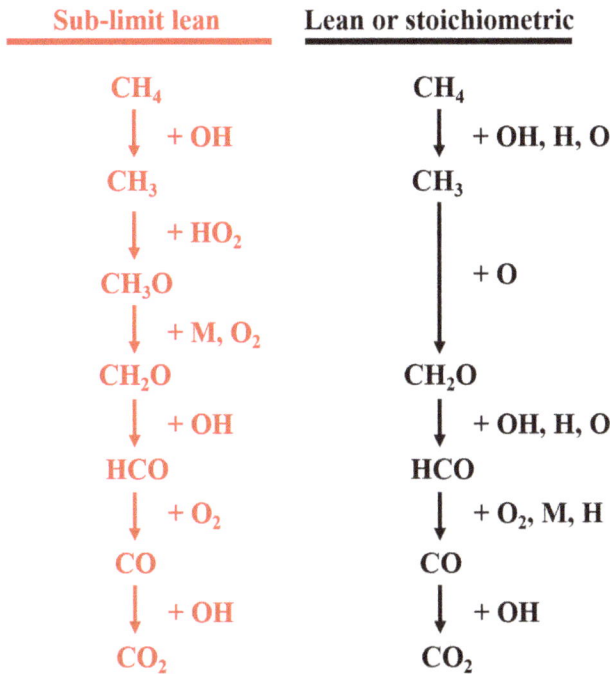

Figure 14. The major reaction pathways of ultra-lean methane combustion as compared to a conventional methane flame.

6. Conclusions

This study successfully demonstrated the feasibility of stably sustaining an ultra-lean premixed methane–air flame in a novel sandwich triple-jet burner, and also successfully explained the structure of the ultra-lean flame in terms of the hydrodynamic field, thermodynamic field, and cross-stream chemical species activities. The simulated results made it possible to explain the stabilization mechanisms in terms of lateral impingement and species exchange routes. It was found that the ultra-lean center flame could be sustained with the support of thermal energy and active radicals diffused from the post-flame of the adjacent lean flames to stabilize the ultra-lean combustion of the central jet. The lateral impingement in the post-flame region accommodated a slow speed of ultra-lean combustion, extending the residence time and allowing chemical species and thermal energy to perform cross-stream diffusion. Remarkably, the methane coming out of the center flame was heated, decomposed, ignited, and burnt with the support of the outer post-flame, which was the stabilization mechanism of the supported ultra-lean methane flames.

Due to the paths of mass and thermal diffusion between the post-flame of the outer flame and the center flame, the central ultra-lean flame can be stabilized and further react while releasing heat. With a greater number of the OH radicals in the post-flame of the outer flame, the OH radical and its participating reactions dominate the initial decomposition and subsequent reactions of methane in the ultra-lean center flame. On the other hand, with the lower flame temperature, the ultra-lean center flame tends to take the slower HO_2 route for further CH_3 oxidation reaction. This work provides the outstanding chemical reaction paths, combustion mechanism, and flame structure of the central ultra-lean flame stabilized in a novel triple-jet burner, which represents the combustion/stabilization mechanism of the supported ultra-lean methane flames.

Author Contributions: H.-C.L. and Y.-C.C. contributed to the concept, numerical simulation, results explanation and writing of the paper. G.-B.C. contributed to the numerical simulation and results explanation. F.-H.W. and H.-Y.L. contributed to paper review and editing.

Funding: This research was funded by the Ministry of Science and Technology of Republic of China grant number NSC 95-2221-E-006-392-MY3. The computer time and CFD package were provided by the National Center for High-Performance Computing, Taiwan, ROC.

Conflicts of Interest: The authors declare no conflicts of interest.

References

1. Montzka, S.A.; Dlugokencky, E.J.; Butler, J.H. Non-CO_2 greenhouse gases and climate change. *Nature* **2011**, *476*, 43–50. [CrossRef] [PubMed]
2. Foster, P.; Ramaswamy, V.; Artexo, P.; Berntsen, T.; Betts, R.; Fahey, D.W. Changes in atmospheric constituents and in radiative forcing. In *Climate 2007: The Physical Science Basis*; Solomon, S., Qin, D., Manning, M., Chen, Z., Marquiz, M., Averyt, K.B., Eds.; Cambridge University Press: Cambridge, UK, 2007; pp. 129–234.
3. Lombardi, L.; Carnevale, E.; Corti, A. Greenhouse effect reduction and energy recovery from waste landfill. *Energy* **2006**, *31*, 3208–3219. [CrossRef]
4. Jiang, X.; Mira, D.; Cluff, D.L. The combustion mitigation of methane as a non-CO_2 greenhouse gas. *Prog. Energy Combust. Sci.* **2018**, *66*, 176–199. [CrossRef]
5. Su, S.; Yu, X. A 25 kWe low concentration methane catalytic combustion gas turbine prototype unit. *Energy* **2015**, *79*, 428–438. [CrossRef]
6. Schuller, T.; Durox, D.; Candel, S. Self-induced combustion oscillations of laminar premixed flames stabilized on annular burners, Combust. *Flame* **2003**, *135*, 525–537. [CrossRef]
7. Lee, B.-J.; Kim, J.-S.; Lee, S. Enhancement of blowout limit by the interaction of multiple nonpremixed jet flames. *Combust. Sci. Technol.* **2004**, *176*, 482–497. [CrossRef]
8. Hawkes, E.R.; Chen, J.H. Direct numerical simulation of hydrogen-enriched lean premixed methane-air flames. *Combust. Flame* **2004**, *138*, 242–258. [CrossRef]
9. Ren, J.-Y.; Egolfopoulos, F.N. NOx Emission control of lean methane-air combustion with addition of methane reforming products. *Combust. Sci. Technol.* **2002**, *174*, 181–205. [CrossRef]
10. Joannon, M.D.E.; Sabia, P.; Tregrossi, A.; Cavaliere, A. Dynamic behavior of methane oxidation in premixed flow reactor. *Combust. Sci. Technol.* **2004**, *176*, 769–783. [CrossRef]
11. Kimura, I.; Ukawa, H. Studies on the Bunsen Flames of Fuel-Rich Mixture. In Proceedings of the 8th Symposium (International) on Combustion, Pasadena, CA, USA, 28 August–3 September 1960; The Combustion Institute: Pittsburgh, PA, USA, 1956; pp. 521–523.
12. Singer, J.M. Burning Velocity Measurements on Slot Burners; Comparison with Cylindrical Burner Determination. In Proceedings of the 4th Symposium (International) on Combustion, Cambridge, MA, USA, 1–5 September 1952; The Combustion Institute: Pittsburgh, PA, USA, 1952; pp. 352–358.
13. Menon, R.; Gollahalli, S.R. *Multiple Jet Gas Flames in Still Air, ASME, HTD*; ASME: New York, NY, USA, 1985; Volume 45, pp. 27–133.
14. Menon, R.; Gollahalli, S.R. Investigation of Interacting Multiple Jets in Cross Flow. *Combust. Sci. Technol.* **1988**, *60*, 375–389. [CrossRef]
15. Roper, F.G. Laminar Diffusion Flame Sizes for Interacting Burners. *Combust. Flame* **1979**, *34*, 19–27. [CrossRef]
16. Seigo, K.; Satoshi, H.; Yoshito, U.; Syuichi, M.; Katsuo, A. Characteristics of combustion of Rich-Lean Flame Burner under Low Load Combustion. In Proceedings of the 20th International Colloquium on the Dynamics of Explosion and Reactive Systems, Montreal, QC, Canada, 31 July–5 August 2005.
17. Lin, H.-C.; Chen, B.-C.; Ho, C.-C.; Chao, Y.-C. A Study of Mutual Interaction of Two Premixed Flame in Lean Combustion. In Proceedings of the Asia-Pacific Conference on Combustion, Nagoya, Japan, 20–23 May 2007.
18. Lin, H.-C.; Chen, B.-C.; Ho, C.-C.; Chao, Y.-C. Mutual Interaction of Two Methane Premixed Flames in Extremely Lean Combustion. In Proceedings of the 21th International Colloquium on the Dynamics of Explosion and Reactive Systems, Poitiers, France, 23–27 July 2007.
19. Chao, B.H.; Egolfopoulos, F.N.; Law, C.K. Structure and Propagation of Premixed Flame in Nozzle-Generated Counterflow. *Combust. Flame* **1997**, *109*, 620–638. [CrossRef]

20. Chelliah, H.K.; Bui-Pham, M.; Seshadri, K.; Law, C.K. Numerical Description of the Structure of Counterflow Heptane-Air Flames Using Detailed and Reduced Chemistry with Comparison to Experiment. In Proceedings of the Twenty-Fourth Symposium (International) on Combustion, Sydney, Australia, 5–10 July 1992; The Combustion Institute: Pittsburgh, PA, USA, 1992; pp. 851–857.

21. Kaiser, C.; Liu, J.-B.; Ronney, P.D. Diffusive-Thermal Instability of Counterflow Flames at Low Lewis Number. In Proceedings of the 38th Aerospace Sciences Meeting & Exhibit, Reno, Nevada, 10–13 January 2000.

22. Sohrab, S.H.; Ye, Z.Y.; Law, C.K. Theory of Interactive Combustion of Counterflow Premixed Flames. *Combust. Sci. Technol.* **1986**, *45*, 27–45. [CrossRef]

23. Vagelopoulos, C.M.; Egolfopoulos, F.N.; Law, K.C. Further Considerations on the Determination of Laminar Flame Speeds with the Counterflow Twin Flame Technique. In Proceedings of the Twenty-Fifth Symposium (International) on Combustion, Irvine, CA, USA, 31 July–5 August 1994; The Combustion Institute: Pittsburgh, PA, USA, 1994; pp. 1341–1347.

24. Sung, C.J.; Liu, J.B.; Law, C.K. On the Scalar Structure of Nonequidiffusive Premixed Flames in Counterflow. *Combust. Flame* **1996**, *106*, 168–183. [CrossRef]

25. Zheng, X.L.; Law, C.K. Ignition of premixed hydrogen/air by heated counterflow under reduced and elevated pressures. *Combust. Flame* **2004**, *136*, 168–179. [CrossRef]

26. Sung, C.J.; Law, C.K. Structural Sensitivity, Response, and Extinction of Diffusion and Premixed Flames in Oscillating Counterflow. *Combust. Flame* **2000**, *123*, 375–388. [CrossRef]

27. Lin, H.-C.; Cheng, T.-S.; Chen, B.-C.; Ho, C.-C.; Chao, Y.-C. A comprehensive study of two interactive parallel premixed methane flames on lean combustion. *Proc. Combust. Inst.* **2009**, *32*, 995–1002. [CrossRef]

28. Cheng, Z.; Pitz, R.; Wehrmeyer, J. Opposed Jet Flames of Very Lean or Rich Premixed Propane-Air Reactants vs. Hot Products. *Combust. Flame* **2002**, *128*, 232–241.

29. Cheng, Z.; Wehrmeyer, J.A.; Pitz, R.W. Lean or ultra-lean stretched planar methane/air flames. *Proc. Combust. Inst.* **2005**, *30*, 285–293. [CrossRef]

30. Cheng, Z.; Pitz, R.; Wehrmeyer, J. Lean and ultralean stretched propane–air counterflow flames. *Combust. Flame* **2006**, *145*, 647–662. [CrossRef]

31. Law, C.K. *Combustion Physics*; Cambridge University Press: Cambridge, UK, 2006.

32. Smooke, M.D. *Reduced Kinetic Mechanisms and Asymptotic Approximations for Methane-Air Flames*; Lecture Note in Physics 384; Springer: Berlin, Germany, 1991.

33. GRI-Mech. Available online: http://combustion.berkeley.edu/gri-mech/version30/text30.html (accessed on 30 July 1999).

energies

MDPI

Article

Combustion Characteristics and NO$_x$ Emission through a Swirling Burner with Adjustable Flaring Angle

Yafei Zhang [1], Rui Luo [2], Yihua Dou [1],* and Qulan Zhou [3],*

[1] Mechanical Engineering College, Xi'an Shiyou University, Xi'an 710065, China; effyzhang@126.com
[2] Xi'an Thermal Power Research Institute Co., Ltd., Xi'an 710032, China; luorui@tpri.com.cn
[3] State Key Laboratory of Multiphase Flow in Power Engineering, Xi'an Jiaotong University, Xi'an 710049, China
* Correspondence: yhdou@vip.sina.com (Y.D.); qlzhou@mail.xjtu.edu.cn (Q.Z.);
   Tel.: +86-029-88382617 (Y.D.); +86-029-82665412 (Q.Z.)

Received: 1 July 2018; Accepted: 18 August 2018; Published: 20 August 2018

Abstract: A swirling burner with a variable inner secondary air (ISA) flaring angle β is proposed and a laboratory scale opposed-firing furnace is built. Temperature distribution and NO$_x$ emission are designedly measured. The combustion characteristics affected by variable β are experimentally evaluated from ignition and burnout data. Meanwhile, NO$_x$ reduction by the variable β is analyzed through emissions measurements. Different inner/outer primary coal-air concentration ratios γ, thermal loads and coal types are considered in this study. Results indicate that β variation provides a new approach to promote ignition and burnout, as well as NO$_x$ emission reduction under conditions of fuel rich/lean combustion and load variation. The recommended β of a swirling burner under different conditions is not always constant. The optimal β_{opt} of the swirling burner under all conditions for different burning performance are summarized in the form of curves, which could provide reference for exquisite combustion adjustment.

Keywords: swirling burner; flaring angle; fuel rich/lean combustion; low load; combustion adjustment

---

1. Introduction

Combustion adjustment technology is still a major interest for coal-based thermal power plants, because coal plays an important role as the primary energy source in the development of countries of large coal reserves, such as China and Japan [1–3]. With increasing demands of energy conservation and emission reduction, great progress has been made in combustion optimization and higher control level of NO$_x$ emission in academic studies [4–7] and engineering applications [8,9]. Because low-NO$_x$ burning can relieve stress on the denitration system downstream of the flue, research on NO$_x$ control by swirling burners is perennially conducted. The operating parameters of swirling burners are focused on combustion organization and low-NO$_x$ burning. Sung et al. [9] studied the effect of secondary air swirl intensity on flame and NO$_x$ reduction. Katzer et al. [10] explored the relationship between burner operating conditions (including variable loads and air distribution) and flame characteristics. Song et al. [11] discussed the impact of inner and outer secondary air distribution in the burner on aerodynamic characteristics in down-fired boiler. Meanwhile, the structure of swirling burners is also a research interest of many investigators. Wang et al. [12] experimentally researched the effect of inner secondary air vane angles of swirl burners for 300 MW down-fired boilers on NO$_x$ reduction, and gave the optimal angle value. Ti et.al. decreased the NO$_x$ emissions of a 600 MW wall-fired boiler through outer secondary air vane angles optimization [13], and numerically studied the effect of varying swirl burner cone length on the ignition and NO$_x$ emissions in a cylindrical furnace [14]. Chen et al. [15]

added a baffle ring to a swirl burner duct outlet to improve the penetration depth of coal/air flow, which obtained favorable results for NO_x control in a 300 MW down-fired boiler. Li et al. [16] discussed the effect of swirl burner outer secondary air vane angle variation on combustion characteristics and NO_x emissions with laboratory-scale and industrial-scale experiments, respectively. Jing et al. [17] also researched the influence of swirl burner outer secondary air vane angles on combustion and NO_x formation in 300 MW wall-fired boilers. Luo et al. [18] proposed a swirl burner with dual-gear rings and double conical flaring based on dual register structure, and experimentally studied the effect on combustion and NO_x emissions in the laboratory. Zhou et al. [19] numerically studied the effect of a Venturi tube and partition annulus in the primary air pipe on temperature and gas species distribution in a furnace. The optimized swirl burner obtained favorable performance with significant NO_x reduction. In conclusion, the studies of operating parameters of burners are mainly focused on swirl intensity, air distribution and ways of feeding fuel, while the focused burner structures mainly consider the swirl vane angle, adding some new parts in the primary air pipe of burners, and retrofitting burner types. However, fewer papers have reported the effect of swirl burner flaring angle, which is a commonly used structure in swirl burners and also significantly impacts the combustion organization of burners [20]. The flaring angles of burner are mostly still fixed at present, and not utilized as an operating parameter for combustion optimization in power plants.

Fuel rich/lean combustion is an important means of reducing NO_x in coal-fired boilers. Song et al. [21] researched the influence of the mass ratio of coal in fuel-rich flow to that in fuel-lean flow in a 600 MW down-fired boiler, and obtained the optimal mass ratio for high burnout and low NO_x emissions. Zeng et al. [22] studied the effect of coal bias distribution on the slagging with swirl burner organizing combustion. Zhou et al. [23] experimentally studied the effect of the block size and particle concentrations in burner primary air pipe on flow characteristics. Li et al. [24] discussed the effects of particle concentration variation in the primary air duct on combustion and NO_x emissions for swirl burners. Chen et al. [25] studied the fuel bias influence in the primary air duct on gas/particle flow characteristics. However, the effects of swirl burner flaring angle variation on rich/lean combustion have not been considered. Moreover, in order to improve the capacity of renewable energy sources, developing deep peaking transformation for coal-fired power plant is meaningful [26]. Some researchers have studied combustion characteristics under different loads, including ignition and NO_x formation [15,27]. Nonetheless, how to use burner flaring for improving ignition under different loads is still less reported in the literature.

In this paper, a swirling burner with real-time adjustable inner secondary air flaring angle is proposed and built on a laboratory scale. Although some related works [20,28] have been conducted about the burner adjustable flaring angle influence mechanism and combined control with air distribution, its application under various load and fuel rich/lean combustion conditions is still not reported. The effects of burner flaring angle on fuel rich/lean combustion and low load combustion are separately studied by combustion experiments in a laboratory-scale opposed-firing furnace. The authors have strived to reveal the burner flaring variation rules for application in combustion adjustment.

2. Experimental Setup and Research Methods

The studied swirling burner is based on a dual register swirling burner structure, which is composed from the inside out of a center pipe, inner/outer primary air (IPA/OPA) pipe and inner/outer secondary air (ISA/OSA) pipe, as Figure 1 shows. The novel flaring of the burner is composed of multiple metal flakelets partly stacked in a circumferential layout, as Figure 1a shows. Thus, the novel flaring angle varies when the flakelets are rotating. Each flakelet is connected with pin on the end of the ISA straight pipe. Each pin is fixed to one linkage and all the linkages are connected by a rear ring. In the experiment, two steel rods are separately fixed on the rear ring symmetrically, so as to vary the flaring angle β by pulling or pushing the steel rods.

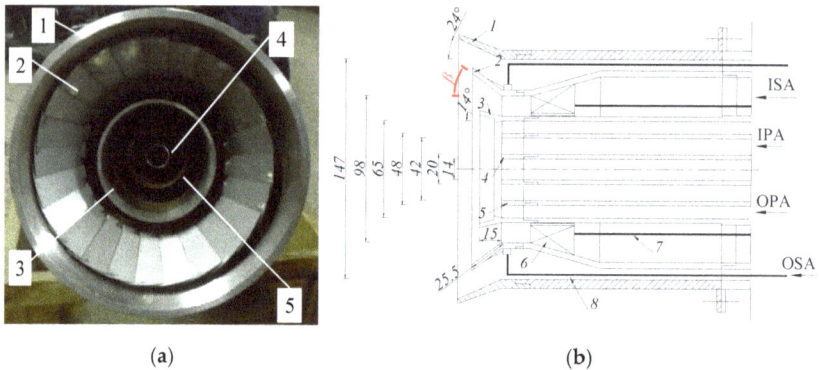

(a) (b)

Figure 1. Schematic diagram of the novel swirling burner: (**a**) Spout photograph of the swirling burner; (**b**) Assembly diagram of this novel burner: (1) outer-secondary air flaring; (2) adjustable inner-secondary air flaring; (3) outer-primary air flaring; (4) center air pipe; (5) inner-primary air pipe; (6) swirling vane; (7) the pull rod of the swirling vane; (8) the pull rod of the adjustable flaring.

A laboratory-scaled furnace combustion test system is set up, which is comprised of a furnace body system, pulverized coal feeding system, air and gas system, water cooling system, ignition system and measurement system, as Figure 2 shows. The furnace body system provides a pulverized coal burning space and simulates the features of actual boilers. The coal feeding system contains four coal feeders and related pipes, which guarantees a persistent fuel supply. The air and gas system provides fresh air for coal burning, and exhausts the flue gas from furnace. The water cooling system serves for cooling the flue gas from the outlet of the furnace and all the measurement equipment working in the furnace. The ignition system uses an oil gun to heat the furnace and ignite the pulverized coal at the beginning of the tests.

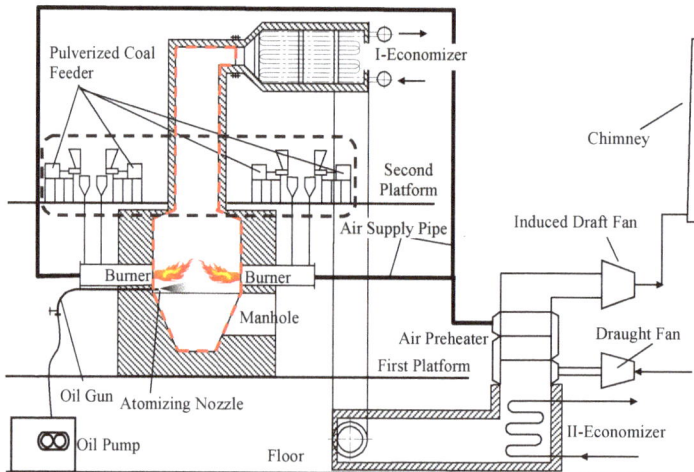

Figure 2. Combustion test system.

A couple of the proposed swirling burners are symmetrically installed on the front and rear walls of the lower furnace. The lower furnace cross-section is 1.0 m × 0.8 m for depth × width. A rectangular coordinate system is built for the furnace, the origin of which is set as the center of the burner couple axis. The dimensionless depth, width and height of furnace are separately defined as $X = x/a$, $Y = y/a$, $Z = z/a$, where a is the axis distance of front and rear wall burner spout as a reference size.

Two different bituminous coals from Huangling County of Shaanxi Province (HL coal) and Wuhai of Inner Mongolia Province (WH coal) in China are separately used in this study. The results of proximate and ultimate analysis, as well as net calorific value, are detailed in Table 1. The coals are pulverized before the combustion tests, and supplied through two coal feeders into the IPA pipe and OPA pipe of one swirl burner for burning in the furnace. Fuel rich/lean combustion is realized through controlling the pulverized coal concentration in the IPA and OPA pipes. The fuel rich/lean ratio (γ) is defined as the ratio of pulverized coal concentration in IPA to that in OPA. The fuel rich/lean ratio (γ) is varied from 1 to 3 in this study through adjusting the coal feeding rate in the inner/outer primary air pipes, respectively. The thermal power input is set to 0.7, 0.6 and 0.5 MW by changing the coal amount of the feeders. The ISA flaring expanding or shrinking can be adjusted in real-time by pulling and pushing the burner tie rod, thus the burner ISA flaring angle (β) changes. All the parameters and study conditions are listed in Table 2.

The swirl number n is calculated using Equation (1) [29,30]:

$$n = \frac{2}{3} \left[\frac{1 - (d_i/d_o)^3}{1 - (d_i/d_o)^2} \right] \tan(\theta) \tag{1}$$

where d_i is the inner diameter of the swirler, and d_o is the outer diameter of the swirler. Thus, the swirl number n primarily depends on the swirl vane angle. Because the swirl vane angle in the ISA pipe is fixed in this study, the swirl number of co-axial jetting is constant.

The measurement system of the experiments is comprised of thermocouples, a gas analyzer and sampling equipment. Before measurements the thermocouple device and flue gas analyzer are calibrated. A water-cooled probe was utilized in the experimental measurements to prevent the equipment from suffering burnout [12]. The flue gas temperature in lower furnace is monitored with water-cooled PtRh10-Pt thermocouples (Xi'an Xiyi Industrial Control Instrument Factory, Xi'an, China). The gas temperature in the upper furnace is measured with NiCr-NiSi thermocouples nested in porcelain sleeves. These two thermocouples are calibrated with a relative error of 0.75% |t|. To avoid any interferences, such as soot pollution and unsteady situations during switching between two different conditions, sufficient time was given between measurements to ensure the accuracy of measurements [15]. For quantitative description and analysis of the combustion parameters, T_{ig} and T_{max} are extracted and compared between the different conditions. The gas temperature of the measuring point close to the spout along the burner axis direction is considered as the ignition characteristic temperature T_{ig}, and the highest gas temperature T_{max} along the furnace height is considered as the combustion intensity.

NO_x emissions are obtained through a GASMET-DX4000 flue gas analyzer (Gasmet Technologies Oy, Helsinki, Finland) with an accuracy of ±2 vol %. The oxygen component of flue gas is monitored by a MSI-Compact flue gas analyzer (Drägerwerk AG & Co., Lübeck, Germany) with an accuracy of ±0.3 vol %. The experimentally obtained NO data were converted to the standard of 6% O_2 according to the Equation (2) [27]:

$$NO_x(\text{ppm @6%}O_2) = \frac{NO(\text{ppm})}{0.95} \times \frac{21 - 6}{21 - O_2(\%)} \tag{2}$$

where NO_x (ppm @6%O_2) in standard state, 6% O_2, (ppm); NO (ppm) is the measured volume fraction of NO, (ppm); O_2 is the volume fraction of oxygen, (%); 0.95 is the assumed ratio of NO to total NO_x.

Unburned carbon in the fly ash is sampled at the furnace outlet using a water-cooled probe as Figure 3 shows. The particles in the sampled fly ash continue burn in the thermal gravimetric analyzer (TGA). The burnout can be calculated by the following formula [28,30–32]:

$$\psi = \frac{[1 - (w_k/w_x)]}{(1 - w_k)} \tag{3}$$

where ψ is char burnout, w is the ash weight fraction, and the subscript k and x refer to the ash contents in the input coal and char sample, respectively.

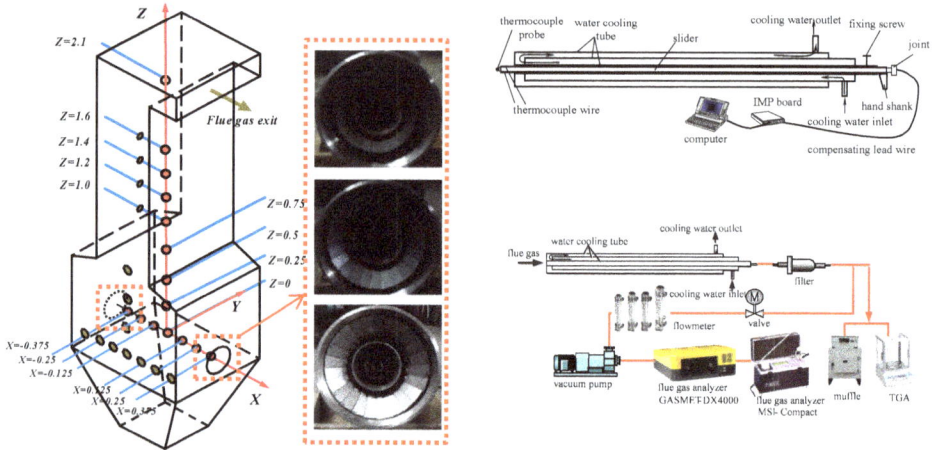

Figure 3. Measurement system scheme.

Table 1. Coal properties (as received).

Fuel	Proximate Analysis/%			Ultimate Analysis/%					NCV (Net Calorific Value)/MJ·kg^{-1}
	Mar	Aar	Vdaf	Car	Har	Nar	Oar	Sar	
HL coal	6.80	13.59	38.00	65.67	3.95	0.85	8.60	0.54	24.93
WH coal	1.39	43.94	32.08	42.87	2.70	0.66	8.01	0.43	16.29

Table 2. Study conditions and concerning parameters.

Num	Parameter	Variable Symbol	Unit	Value
1	Coal	-	-	HL coal, WH coal
2	Ratio of rich(inner) to lean(outer) for pulverized coal concentration	γ	-	1, 2, 3
3	Thermal load	Q	MW	0.7, 0.6, 0.5
4	Swirl number	n	-	$n_1 = 0$, $n_2 = 0$, $n_3 = 0.95$, $n_4 = 0$
5	OPA flaring angle	β_{OPA}	°	14
6	ISA flaring angle	β	°	11.4, 17.1, 26.0, 31.7, 35.5
7	OSA flaring angle	β_{OSA}	°	24
8	Air temperature	T_0	K	343
9	Excess air coefficient	α	-	1.2
10	Air ratio	V	-	$V_1 = 0.1$, $V_2 = 0.1$, $V_3 = 0.6$, $V_4 = 0.2$

3. Results and Discussion

3.1. Effect on Fuel Rich/Lean Combustion

Coal bias burning is an important factor of low nitrogen burning. In the following analysis, the effect of β on bias burning is researched under a thermal load of 0.7 MW. Figure 4 shows the gas temperature along the furnace depth and height direction under a certain condition, respectively. Figure 4a shows that the gas temperature rises from burner spout to the furnace center, which reflects the pulverized coal ignition process. The average of two measurement points of each curve close to burner spout is considered as T_{ig} for a specific condition. Figure 4b displays that the gas temperature of the lower furnace rises with fluctuation along the furnace centerline, and gradually decreases in the upper furnace in the burnout stage. The T_{max} is extracted from the maximum value of one temperature distribution curve for each condition.

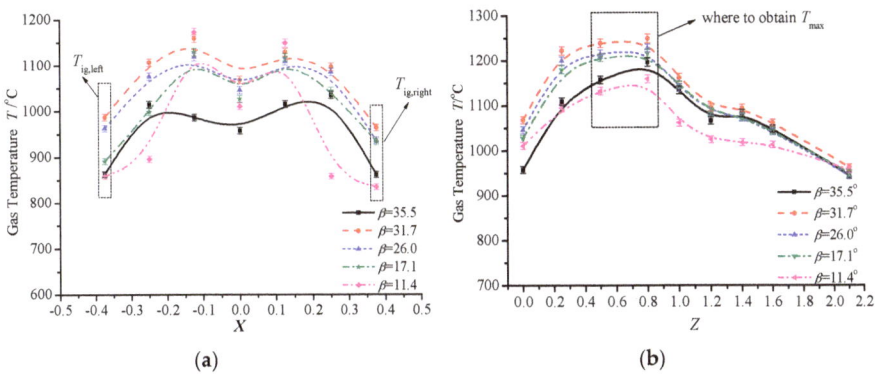

Figure 4. Temperature distribution in test furnace (Huangling County of Shaanxi Province coal (HL coal) burning, $\gamma = 2$, $Q = 0.7$ MW): (a) along the depth direction; (b) along the height direction.

Adjustment of swirl burner ISA flaring angle β affects the temperature distribution along the furnace depth. Enlarging the flaring angle (β = from 11.4° to 31.7°) improves the reverse flow zone to induce hot gas closer to spout, while the gas temperature in the furnace center is not high. However, too large a flaring angle ($\beta = 35.5°$) of the burner maybe leads to open airflow outside the burner spout, which is detrimental to ignition. In Figure 4a, the condition of $\beta = 26.0°$–31.7° favors ignition improvement. In some case, β could be adjusted smaller to prevent the burner spout from burnout.

Figure 4b shows that the gas temperature along the furnace centerline rises from burner layer to the outlet of the lower furnace, and then decreases gradually in the upper furnace. The temperature level in the lower furnace is significantly affected by the burner flaring angle β. The condition of larger β of burner ($\beta = 31.7°$) can achieve a higher temperature level of the furnace in Figure 4.

Figure 5 summarizes T_{ig} under conditions of different γ for bias burning. T_{ig} of HL coal appears above 750 °C because of this coal's high calorific value and volatile content. When $\gamma = 1$, the condition of $\beta = 35.5°$ displays a better ignition performance, because a large β improves the reverse hot flow to the spout significantly. When β becomes smaller, the temperature distribution along the burner axis appears higher in the center and lower near the spout. If bias burning works, the temperature along the burner axis appears higher at center and lower aside as well. When $\gamma = 2$, the condition of $\beta = 26.0°$–31.7° is in favor of ignition improvement. When $\gamma = 3$, the condition of large β loses the ability of helping coal ignition. The swirling burner with smaller β ($\beta = 17.1°$) ignites coal better under condition of higher γ.

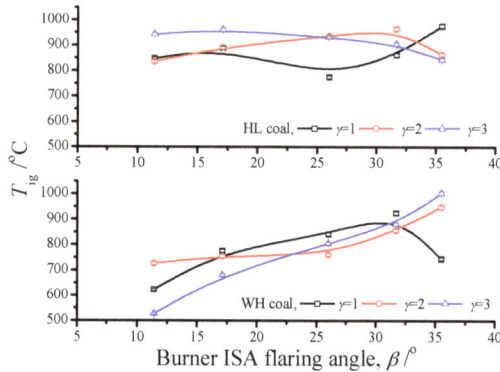

Figure 5. Ignition characteristic temperature variation with different β and γ.

The ignition of WH coal is worse than HL coal, thus the impact of γ and β is more significant. Under different γ conditions, increasing β generally improves ignition. WH coal needs more ignition heat and therefore requires a larger flaring angle to form a high temperature gas reflux. When dense-dilute burning works (γ is larger), the flaring angle β enlargement of the burner benefits ignition and temperature level along the burner axis. This indicates that when the inner primary airstream has high coal concentration, a swirl burner with larger β can draw hot gas near the burner spout to ignite the thick coal stream, and improve the temperature level near the burner spout rapidly.

Figure 6 compares the highest temperature level of the furnace. On conditions of HL coal combustion, T_{max} in the furnace seems more affected by γ than β. Because of the flammable characteristics of HL coal, its combustion temperature depends more on fuel concentration than airflow variation slightly by β. Fuel rich/lean ratio γ directly affects local fuel concentration distribution, so as to influence temperature level more significantly. When non-bias or low bias burning applies, the coal can burn strongly because of the reasonable fuel-air mixing ratio. The variation of β affects the temperature level in some extent through adjusting the inner secondary airflow. When the fuel rich/lean ratio γ rises, coal burning worsens. The secondary airflow has less impact than the primary airflow. The impact of β on the gas temperature level appears insignificant. Therefore, Figure 6 shows that burners with $\beta = 26.0°-31.7°$ achieve higher temperature levels in the furnace for lower γ, while conditions of different β give almost the same temperature level for higher γ.

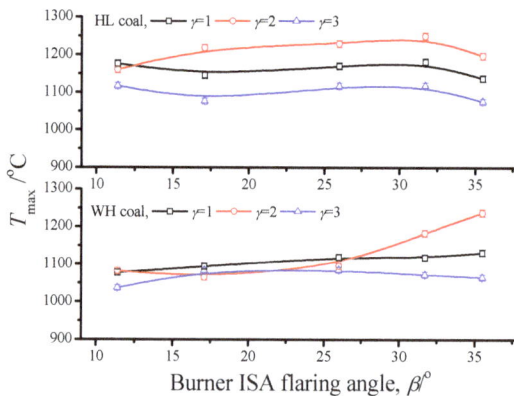

Figure 6. Maximum temperature variation with different β and γ.

Conditions of WH coal burning are more affected by β than γ. WH coal is less flammable than HL coal, thus its combustion temperature level is more affected by the airflow of the burner. When fuel rich/lean ratio is in a median range, a larger β obviously promotes T_{max} as the condition of $\gamma = 2$ shows. This states that under this fuel concentration distribution situation, a burner with $\beta = 35.5°$ would provide the most appropriate air supply mode for combustion temperature level promotion. When non-bias burning or high-bias burning works, the β variation has limited impact on T_{max}. In this situation, fuel supply plays a leading role in the temperature level and weakens the secondary air function. For example, burners with $\beta = 26°$ provide a little higher temperature level under condition of $\gamma = 1$ and $\gamma = 3$. Thus, combustion adjustment through both γ and β are necessary for lean coal.

Figure 7 displays the NO_x emissions and burnout with two different coals, separately. NO_x reduction is generally realized through low-NO_x burning in furnace or gas denitration after burning. The previous technology has been popularly utilized for fuel rich/lean combustion, which controls NO_x formation through restricting the local low oxygen concentration in the high temperature zone. Fuel rich/lean combustion can basically reduce NO_x emissions significantly, as the hollow-dot solid lines in Figure 7 show. Moreover, NO_x emission reduction can also be realized through burner flaring angle β adjustment for both HL coal and WH coal. That is because the variation of flaring angle β changes the secondary airflow direction and jetting rigidity, so as to influence fuel and air mixing outside the burner spout. For example, if the inner secondary air flaring angle is enlarged, the outlet area of the inner secondary air increases, and the rigidity of airflow decreases. Meanwhile, the outer secondary airflow becomes more rigid, and the flow direction moves away from the burner axis. Thus, the mixing between pulverized coals with different layer airflow becomes different. It can also be considered as local rich/lean combustion in another way. When fuel rich/lean ratio γ keeps constant, variations of β affect NO_x emission more significantly for non-rich/lean combustion than that for fuel rich/lean combustion. It is inferred that local rich/lean combustion caused by flaring angle β variation plays a more important role in non-rich/lean combustion for NO_x reduction. If fuel rich/lean combustion employs a swirling burner, less effect of NO_x formation control is obtained through the "local" fuel rich/lean combustion caused by β variation. In addition, the NO_x emission of condition $\beta = 35.5°$ reached a much lower level than other conditions, which is contributed by both extra local rich/lean combustion and combustion deterioration at the cost of low burnout.

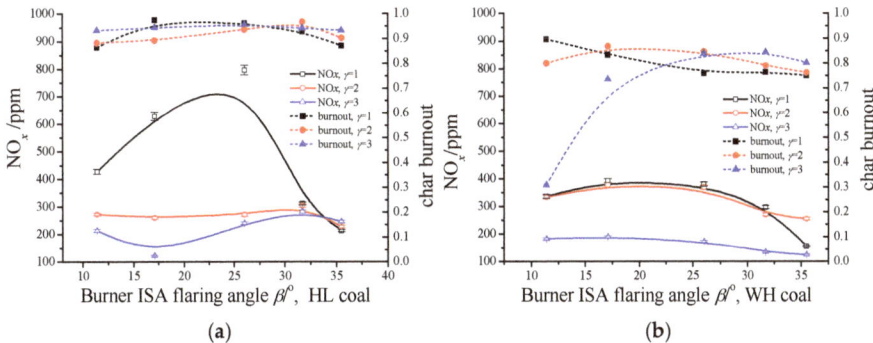

Figure 7. NO_x emission and burnout comparison with different β and γ: (**a**) HL coal; (**b**) Wuhai of Inner Mongolia Province coal (WH coal).

The optimal β for NO_x control is not the same for the two different coals. For example, under condition of $\gamma = 3$, the suggested β is 17.1° for HL coal but 35.5° for WH coal. HL coal ignition is better than WH coal as Table 1 shows. It can be inferred that a more reducing atmosphere forms in the outside spout for HL coal than WH coal. A smaller β is enough for HL coal to control NO_x formation than that needed for WH coal. Variation of β brings more adaptability for burners. That's also the reason that the burner with real-time adjustable flaring angle is proposed and studied in this paper.

Burnout is impacted by both fuel rich/lean ratio and burner flaring angle in this study, and illustrated in Figure 7. The effect of burner flaring angle β is determined through changing the secondary airflow and fuel-air mixing in a later period, which seems more significant than that of fuel rich/lean ratio γ for all conditions. When burning difficult-flammable WH coal, the smaller β in non-rich/lean combustion and larger β in fuel rich/lean combustion are suggested for high burnout. Whatever coal is used, HL coal or WH coal, the burner flaring angle β can be considered as an auxiliary adjustment to control NO_x emissions and further improve burnout.

3.2. Effect on Load Variation

To study the effect of β on ignition and stable combustion under variable loads, the characteristic parameters (T_{ig} and T_{max}) are extracted from the measured temperature distributions for analysis as Figure 8 shows. Ignition characteristic temperature is mainly decided by the thermal load for both HL coal and WH coal, because the thermal load influences the global temperature level in the furnace. Moreover, ignition is also affected by the hot flue gas entrainment ability, which can be adjusted by the inner secondary air flaring angle β.

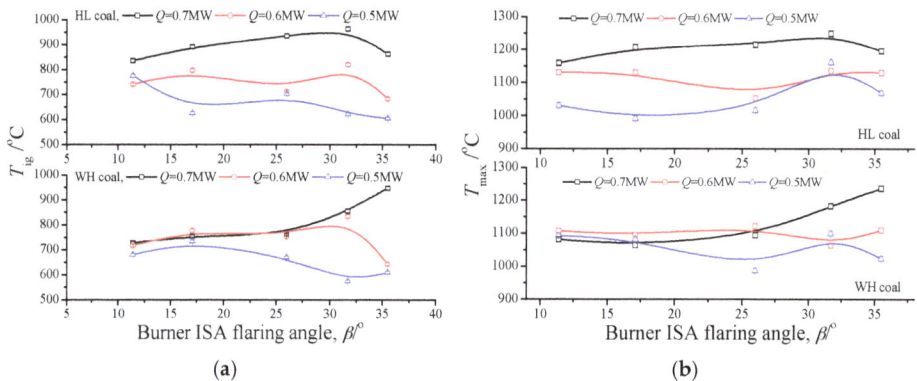

Figure 8. Ignition comparison among the conditions of different β and loads: (**a**) Ignition characteristics; (**b**) combustion intensity characteristics.

High thermal load could not ensure a high flame temperature level, because an increasing load requires more fresh air supply. Fresh cold air would possibly decrease the temperature level in the furnace if pulverized coal does not ignite and release heat in a timely way in the combustion process. Therefore, the highest temperature value in the furnace depends on the mixing of pulverized coal and air, which can be realized by the timely flaring angle variation of the proposed swirling burner.

In Figure 8a the enlargement of β does not always help improving HL coal ignition. Especially when the gas temperature level is low, a large β results in the cold reflux flue gas which does not benefit ignition. The combustible HL coal ignition needs fresh air more than temperature. Thus, under low thermal load with burning HL coal, a small burner β promotes primary/secondary air mixing earlier which could better improve ignition.

For difficult-flammable WH coal, load increasing improves WH ignition significantly. Under low thermal load, burners with $\beta = 11.4°–26.0°$ provide the best stable ignition. This suggests that low load conditions require a small burner β for ignition. The β for WH coal ignition appears larger than that for HL coal. That's because difficult-flammable WH coal ignition requires not only oxygen, but also hot flue gas in order to reach a suitable reactivity level.

The effects on temperature level of the furnace under variant loads were also researched. Figure 8b shows that T_{max} is mainly impacted by the thermal load, especially for HL coal. Under low load β is suggested to rise to above 31.7° for HL coal, while reducing β to below 17.1° for WH coal, which would benefit timely ignition and strong burning. The inferred explanation is that burners with small β can adapt to the situation of coal-air supply reduction, and avoid forming open airflows. In a word, combustion adjustment with β promotes stable burning, especially under conditions of low thermal load.

For example, whether burning HL coal or WH coal, the difference of characteristic ignition temperature between various loads conditions becomes smaller if we adjust the burner ISA flaring angle to a smaller value ($\beta = 11.4$–$26°$). The temperature difference between $Q = 0.7$ MW and $Q = 0.5$ MW is about 100 °C with $\beta = 11.4°$, but more than 250 °C with $\beta = 35.5°$. Therefore, under higher load the burner flaring angle could be set larger ($\beta = 26.0°$–$35.5°$) for a higher ignition characteristic temperature, but should be set smaller as $\beta = 11.4°$–$17.1°$ for a stable ignition characteristic temperature (the temperature drops less than 200 °C in the combustion experiment). However, there is still a void in using the adjustable flaring angle burner in a practical utility boiler. This experimental study provides some demonstration for further applications.

Figure 9 shows the NO_x emissions and burnouts of different load conditions. Whether HL coal or WH coal is used, the NO_x emission maximum increases with rising thermal load. That is because a higher temperature level in furnace caused by high thermal load leads to more NO_x formation in the combustion. Moreover, varying β could change the NO_x emissions in some extent. For example, burners with larger or smaller β could control combustion with lower NO_x emissions.

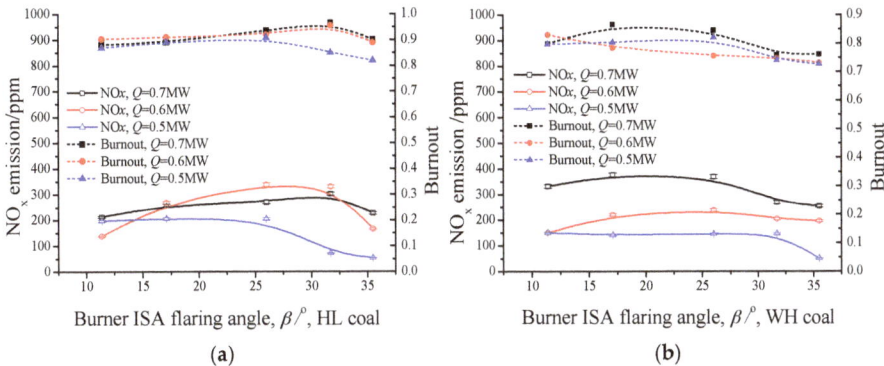

Figure 9. NO_x emission and burnout comparison with different β and Q: (a) HL coal; (b) WH coal.

Burnout is also achieved as shown in Figure 9. Generally, HL coal burning performs higher burnout than WH coal, which is consistent with the inherent fuel characteristics. The burnout is affected by both the thermal load and burner flaring angle. There is an optimal ISA flaring angle β_{opt} corresponding to the highest burnout. β_{opt} decreases with thermal load reduction for both coals. $\beta_{opt} = 26.0°$–$31.7°$ is suggested for HL coal burning, and $\beta_{opt} = 11.4°$–$26.0°$ is suggested for WH coal.

3.3. Combustion Adjustment Suggestion

Combined with an adjustable swirling burner flare angle, combustion can be more improved and optimized through use of a swirling burner. The variation of β affects combustion through swirling burner flaring, which not only guides the inside but also outside airflow from the spout.

For fuel rich/lean combustion and low-load combustion, an adjustable burner flaring angle brings adaptability for coal and load, enriches the combustion optimization methods and offers better potential of clean and efficiency combustion. To expediently guide engineering applications of swirling burner with adjustable flaring angles, the optimal β values corresponding to the condition of easiest ignition, highest burnout and lowest NO_x emission are summarized in Figure 10, respectively.

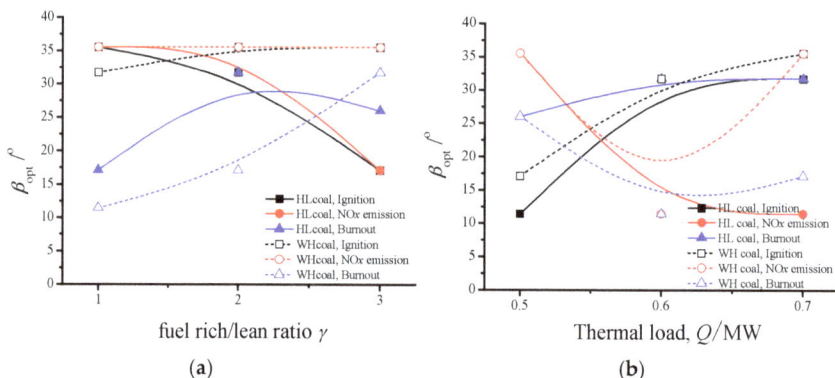

Figure 10. Suggested β adjustment for γ and Q variation: (**a**) for fuel rich/lean combustion; (**b**) for variant-load combustion.

The optimal β of a burner varies according to the operating conditions and the specifics concerning combustion performance. The variation rules for diverse coals, fuel rich/lean ratios and thermal loads are different. For example, when the fuel rich/lean ratio γ increases, the flaring angle β is suggested to augment for HL coal but is lower for WH coal to ignite coal in a timely way as Figure 10a shows, whereas, when the thermal loads increases, the β is suggested to be raised for both HL and WH coal. Under some conditions, β variation affects combustion adjustment less. For example, NO_x emissions are less impacted by β with burning WH coal when rich/lean ratio varies. Under some other conditions, the effects of β display parabolic curve rules, such as NO_x emission and burnout of WH coal with load variations. The flaring angle β variation enriches the necessary adjustment methods and flexibility for combustion optimization in engineering applications.

Because the adjustable flaring angle of the burner is composed of rotating multi-flakelets instead of traditional fixed geometry flaring, it is possible to install the adjustable flaring burners on an actual boiler, and control the flaring angle through some mechanical structures, such a pullrod. It is also necessary to apply this burner for solving stable ignition problems under low-load and fuel rich/lean combustion conditions. The combustion experiments are designed and operated according to the similarity rules of experimental fluid. The fuel characteristics, chemical reactions and heat transfer processes are almost the same as the actual conditions. The qualitative results from this work can be directly referred to for actual use. For example, if the fuel rich/lean combustion with this adjustable flaring burner is conducted in a boiler burning WH coal, when the fuel rich/lean ratio needs to augment for lower NO_x formation, the burnout may worsen. In this situation, Figure 10 tells us that the flaring angle should simultaneously be adjusted to a larger value, which benefits WH coal burnout, as the hollow blue triangle line shows in Figure 10.

4. Conclusions

The effect of adjustable flaring of swirling burner on fuel rich/lean combustion and variant load combustion was investigated through combustion experiments in a laboratory-scale furnace. The evaluation of ignition, NO_x emission and burnout of each test conditions indicates that the ISA flaring angle of burner should adapt different combustion conditions to adjust instead of keeping it fixed, which is necessary for combustion optimization. The primary results can be summarized as follows:

(1) Under fuel rich/lean combustion conditions, burner ISA β variation could promote ignition characteristic temperatures above 200 °C. For ignition improvement it is suggested to reduce β for HL coal but to augment it for WH coal when rich/lean ratio γ increases. NO_x emissions are less affected by β than γ, but variation of β has a further reduction effect on NO_x emissions than fixed β. Rising β could promote burnout for both coals if γ increases for fuel rich/lean combustion.

(2) Under fuel rich/lean combustion conditions, burner ISA β diminution could promote ignition for both coals when load decreases. Under the same load, the optimal β obtains about 50–100 °C higher ignition characteristic temperature than the worst β condition. Variation of β can reduce NO_x emissions by about 50 ppm and enhance burnout about 10% compared with the worst β conditions under the same load.

In conclusion, the optimal β of a burner should vary according to the operating conditions and the specific concerning combustion performance. Finally the suggested β for ignition, NO_x emission and burnout with variation of rich/lean ratio and thermal load is summarized as curve group for engineering reference. The detailed rules for more coal types should to be researched in future work.

Author Contributions: Conceptualization, R.L. and Q.Z.; Formal analysis, Y.Z.; Funding acquisition, Y.Z., Y.D. and Q.Z.; Investigation, Y.Z. and R.L.; Methodology, Y.Z. and R.L.; Supervision, Q.Z.; Writing—original draft, R.L.; Writing—review & editing, Y.Z. and Y.D.

Funding: This research was funded by National Science and Technology Major Project of the Ministry of Science and Technology of China (grant number: 2016ZX05017-006-HZ03); Natural Science Foundation of Shaanxi Province of China (grant number: 2017JQ5108); Special Scientific Research Plan of Shaanxi Province Education Department (grant number: 17JK0594); Ministry of Industry and Information Technology Support Project for High-Tech Ships and the Fundamental Research Funds for the Central Universities in Xi'an Jiaotong University.

Nomenclature

Variables

a	depth of furnace (m)
x, y, z	depth, width and height coordinate of the furnace (m)
X, Y, Z	dimensionless of depth, width and height of the furnace (-)
V	air ratio
n	swirl number
α	excess air ratio
β	inner secondary air flaring angle (°)
w	mass fraction (%)
ψ	char burnout
Q	thermal load
γ	ratio of rich(inner) to lean(outer) for pulverized coal concentration
T	temperature

Abbreviations

IPA	inner primary air
OPA	outer primary air
ISA	inner secondary air
OSA	outer secondary air

Subscripts

1/2, 3/4	inner/outer primary air, inner/outer secondary air
k	input coal
x	char sample
opt	optimization
ig	ignition
max	maximum

References

1. Chang, S.; Zhuo, J.; Meng, S.; Qin, S.; Yao, Q. Clean coal technologies in China: Current status and future perspectives. *Engineering* **2016**, *2*, 447–459. [CrossRef]
2. Guan, G. Clean coal technologies in Japan: A review. *Chin. J. Chem. Eng.* **2017**, *25*, 689–697. [CrossRef]
3. Zhang, L.; He, C.; Yang, A.; Yang, Q.; Han, J. Modeling and implication of coal physical input-output table in China—Based on clean coal concept. *Resour. Conserv. Recycl.* **2018**, *129*, 355–365. [CrossRef]
4. Gaikwad, P.; Kulkarni, H.; Sreedhara, S. Simplified numerical modelling of oxy-fuel combustion of pulverized coal in a swirl burner. *Appl. Therm. Eng.* **2017**, *124*, 734–745. [CrossRef]
5. Fan, W.; Li, Y.; Guo, Q.; Chen, C.; Wang, Y. Coal-nitrogen release and nox evolution in the oxidant-staged combustion of coal. *Energy* **2017**, *125*, 417–426. [CrossRef]
6. Xiouris, C.Z.; Koutmos, P. Fluid dynamics modeling of a stratified disk burner in swirl co-flow. *Appl. Therm. Eng.* **2012**, *35*, 60–70. [CrossRef]
7. Sanmiguel-Rojas, E.; Burgos, M.A.; del Pino, C.; Fernandez-Feria, R. Three-dimensional structure of confined swirling jets at moderately large reynolds numbers. *Phys. Fluids* **2008**, *20*, 044104. [CrossRef]
8. Kryjak, M.; Dennis, J.; Ridler, G. Nox reduction using advanced techniques in a 175mwth multi-fuel corner-fired boiler. *Energy Procedia* **2017**, *120*, 689–696. [CrossRef]
9. Sung, Y.; Lee, S.; Eom, S.; Moon, C.; Ahn, S.; Choi, G.; Kim, D. Optical non-intrusive measurements of internal recirculation zone of pulverized coal swirling flames with secondary swirl intensity. *Energy* **2016**, *103*, 61–74. [CrossRef]
10. Katzer, C.; Babul, K.; Klatt, M.; Krautz, H.J. Quantitative and qualitative relationship between swirl burner operatingconditions and pulverized coal flame length. *Fuel Process. Technol.* **2017**, *156*, 138–155. [CrossRef]
11. Song, M.; Zeng, L.; Chen, Z.; Li, Z.; Kuang, M. Aerodynamic characteristics of a 350-mwe supercritical utility boiler with multi-injection and multi-staging: Effects of the inner and outer secondary air distribution in the burner. *J. Energy Inst.* **2018**, *91*, 65–74. [CrossRef]
12. Wang, Q.; Chen, Z.; Che, M.; Zeng, L.; Li, Z.; Song, M. Effect of different inner secondary-air vane angles on combustion characteristics of primary combustion zone for a down-fired 300-mwe utility boiler with overfire air. *Appl. Energy* **2016**, *182*, 29–38. [CrossRef]
13. Ti, S.; Chen, Z.; Li, Z.; Min, K.; Zhu, Q.; Chen, L.; Wang, Z. Effect of outer secondary air vane angles on combustion characteristics and nox emissions for centrally fuel rich swirl burner in a 600-mwe wall-fired pulverized-coal utility boiler. *Appl. Therm. Eng.* **2017**, *125*, 951–962. [CrossRef]
14. Ti, S.; Chen, Z.; Kuang, M.; Li, Z.; Zhu, Q.; Zhang, H.; Wang, Z.; Xu, G. Numerical simulation of the combustion characteristics and no x emission of a swirl burner: Influence of the structure of the burner outlet. *Appl. Therm. Eng.* **2016**, *104*, 565–576. [CrossRef]
15. Chen, Z.; Wang, Q.; Wang, B.; Zeng, L.; Che, M.; Zhang, X.; Li, Z. Anthracite combustion characteristics and no x formation of a 300 mw e down-fired boiler with swirl burners at different loads after the implementation of a new combustion system. *Appl. Energy* **2017**, *189*, 133–141. [CrossRef]
16. Li, S.; Chen, Z.; Li, X.; Jiang, B.; Li, Z.; Sun, R.; Zhu, Q.; Zhang, X. Effect of outer secondary-air vane angle on the flow and combustion characteristics and nox formation of the swirl burner in a 300-mw low-volatile coal-fired boiler with deep air staging. *J. Energy Inst.* **2017**, *90*, 239–256. [CrossRef]
17. Jing, J.; Li, Z.; Liu, G.; Chen, Z.; Ren, F. Influence of different outer secondary air vane angles on flow and combustion characteristics and nox emissions of a new swirl coal burner. *Energy Fuels* **2010**, *24*, 346–354. [CrossRef]
18. Luo, R.; Zhang, Y.; Li, N.; Zhou, Q.; Sun, P. Experimental study on flow and combustion characteristic of a novel swirling burner based on dual register structure for pulverized coal combustion. *Exp. Therm. Fluid Sci.* **2014**, *54*, 136–150. [CrossRef]
19. Zhou, H.; Yang, Y.; Liu, H.; Hang, Q. Numerical simulation of the combustion characteristics of a low nox swirl burner: Influence of the primary air pipe. *Fuel* **2014**, *130*, 168–176. [CrossRef]
20. Luo, R.; Li, N.; Zhang, Y.; Wang, D.; Liu, T.; Zhou, Q.; Chen, X. Effect of the adjustable inner secondary air-flaring angle of swirl burner on coal-opposed combustion. *J. Energy Eng.* **2016**, *142*, 04015018. [CrossRef]
21. Song, M.; Zeng, L.; Yang, X.; Chen, Z.; Li, Z. Influence of the mass ratio of pulverized-coal in fuel-rich flow to that in fuel-lean flow on the gas/particle flow and particle distribution characteristics in a 600 mwe down-fired boiler. *Exp. Therm. Fluid Sci.* **2018**, *91*, 363–373. [CrossRef]

22. Zeng, L.; Li, Z.; Cui, H.; Zhang, F.; Chen, Z.; Zhao, G. Effect of the fuel bias distribution in the primary air nozzle on the slagging near a swirl coal burner throat. *Energy Fuels* **2009**, *23*, 4893–4899. [CrossRef]

23. Zhou, H.; Ma, W.; Zhao, K.; Yang, Y.; Qiu, K. Experimental investigation on the flow characteristics of rice husk in a fuel-rich/lean burner. *Fuel* **2016**, *164*, 1–10. [CrossRef]

24. Li, Z.; Li, S.; Zhu, Q.; Zhang, X.; Li, G.; Liu, Y.; Chen, Z.; Wu, J. Effects of particle concentration variation in the primary air duct on combustion characteristics and nox emissions in a 0.5-mw test facility with pulverized coal swirl burners. *Appl. Therm. Eng.* **2014**, *73*, 859–868. [CrossRef]

25. Chen, Z.; Li, Z.; Jing, J.; Wang, F.; Chen, L.; Wu, S. The influence of fuel bias in the primary air duct on the gas/particle flow characteristics near the swirl burner region. *Fuel Process. Technol.* **2008**, *89*, 958–965. [CrossRef]

26. Gu, Y.; Xu, J.; Chen, D.; Wang, Z.; Li, Q. Overall review of peak shaving for coal-fired power units in China. *Renew. Sustain. Energy Rev.* **2016**, *54*, 723–731. [CrossRef]

27. Li, S.; Chen, Z.; He, E.; Jiang, B.; Li, Z.; Wang, Q. Combustion characteristics and no x formation of a retrofitted low-volatile coal-fired 330 mw utility boiler under various loads with deep-air-staging. *Appl. Therm. Eng.* **2017**, *110*, 223–233. [CrossRef]

28. Luo, R.; Fu, J.; Li, N.; Zhang, Y.; Zhou, Q. Combined control of secondary air flaring angle of burner and air distribution for opposed-firing coal combustion. *Appl. Therm. Eng.* **2015**, *79*, 44–53. [CrossRef]

29. Beér, J.M.; Chomiak, J.; Smoot, L.D. Fluid dynamics of coal combustion: A review. *Prog. Energy Combust. Sci.* **1984**, *10*, 177–208. [CrossRef]

30. Sung, Y.; Choi, G. Non-intrusive optical diagnostics of co- and counter-swirling flames in a dual swirl pulverized coal combustion burner. *Fuel* **2016**, *174*, 76–88. [CrossRef]

31. Li, Z.; Jing, J.; Chen, Z.; Ren, F.; Xu, B.; Wei, H.; Ge, Z. Combustion characteristics and no x emissions of two kinds of swirl burners in a 300-mwe wall-fired pulverized-coal utility boiler. *Combust. Sci. Technol.* **2008**, *180*, 1370–1394. [CrossRef]

32. Sung, Y.; Moon, C.; Eom, S.; Choi, G.; Kim, D. Coal-particle size effects on no reduction and burnout characteristics with air-staged combustion in a pulverized coal-fired furnace. *Fuel* **2016**, *182*, 558–567. [CrossRef]

energies

MDPI

Article

Density Functional Theory Study on Mechanism of Mercury Removal by CeO$_2$ Modified Activated Carbon

Li Zhao, Yang-wen Wu, Jian Han, Han-xiao Wang, Ding-jia Liu, Qiang Lu * and Yong-ping Yang

National Engineering Laboratory for Biomass Power Generation Equipment, North China Electric Power University, Beijing 102206, China; Zhaoli9533@163.com (L.Z.); zorowuyangwen@163.com (Y.-w.W.); hj19940929@163.com (J.H.); jshawanghanxiao@126.com (H.-x.W.); zhaxila@mail.ustc.edu.cn (D.-j.L.); yyp@ncepu.edu.cn (Y.-p.Y.)
* Correspondence: qianglu@mail.ustc.edu.cn; Tel.: +86-010-6177-2030

Received: 28 September 2018; Accepted: 22 October 2018; Published: 23 October 2018

Abstract: Doping of CeO$_2$ on activated carbon (AC) can promote its performance for mercury abatement in flue gas, while the Hg^0 removal mechanism on the AC surface has been rarely reported. In this research, density functional theory (DFT) calculations were implemented to unveil the mechanism of mercury removal on plain AC and CeO$_2$ modified AC (CeO$_2$-AC) sorbents. Calculation results indicate that Hg^0, HCl, HgCl and HgCl$_2$ are all chemisorbed on the adsorbent. Strong interaction and charge transfer are shown by partial density of states (PDOS) analysis of the Hg^0 adsorption configuration. HCl, HgCl and HgCl$_2$ can be dissociatively adsorbed on the AC model and subsequently generate HgCl or HgCl$_2$ released to the gas phase. The adsorption energies of HgCl and HgCl$_2$ on the CeO$_2$-AC model are relatively high, indicating a great capacity for removing HgCl and HgCl$_2$ in flue gas. DFT calculations suggest that AC sorbents exhibit a certain catalytic effect on mercury oxidation, the doping of CeO$_2$ enhances the catalytic ability of Hg^0 oxidation on the AC surface and the reactions follow the Langmuir–Hinshelwood mechanism.

Keywords: flue gas mercury removal; activated carbon sorbent; CeO$_2$ doping; density functional theory(DFT) calculations

---

1. Introduction

Mercury pollution has attracted widespread attention due to the toxic effect, mobility, persistence, and bioaccumulation [1]. In 2013, the Minamata Convention was concluded among more than 90 countries to control mercury emission, which meant the arrival of a new stage in global mercury pollution control [2,3]. Flue gas is regarded as one of the major sources of mercury emitted to the atmosphere. In China, the Emission Standard of Air Pollutants for Thermal Power Plant (GB13223-2011) released in 2011, prescribes the emission of mercury species from coal-fired boilers as below 0.03 mg/m^3 [4–6]. Mercury mainly exists in flue gas as elemental form (Hg^0), oxidized form (Hg^{2+}), and particulate form (Hg^P) [7]. Hg^{2+} and Hg^P can be removed by existing pollutant control equipment in coal-fired power stations, while Hg^0 is hard to control since it is chemically stable and insoluble in water [8–10]. Consequently, Hg^0 removal from flue gas is a research hotspot for mercury pollution control in coal-fired power stations.

Activated carbon injection (ACI) is regarded as a mature technology to control Hg^0 in flue gas, but the high cost makes it insupportable for coal-fired power plants [11,12]. Modification of carbon based sorbents can enhance the adsorption capacity for Hg^0 in flue gas and reduce the consumption of AC sorbents, saving a large amount of operating expenses for coal-fired power plants. Therefore, various modification methods on AC sorbent have been studied recently. CeO$_2$ is a promising catalytic oxidation material which is widely used in the field of denitrification and mercury removal [13–16]. Experimental studies have shown that the doping of metal oxides such as CeO_2 can change the physical and chemical conditions on the AC surface and significantly enhance its mercury removal efficiency in coal-fired flue gas [17–23]. Tian et al. tested the mercury removal performance of CeO_2 doped AC in a simulated flue gas efficiency test and characterization results showed that CeO_2 had good dispersibility on the AC surface and significantly improved the mercury removal efficiency by introducing a large number of functional groups which can be helpful for Hg^0 oxidation [18]. Zhang et al. investigated the mercury removal performance in simulated flue gas by MnO_x and CeO_2 modified semi-coke. It was found that although the pore structure of semi-coke was slightly reduced after loading metal oxide, the mercury removal of semi-coke adsorbent was significantly improved, which was consistent with Wu's research on modification of AC by MnO_x and CeO_2 [22,23].

However, it is difficult to understand how the surface condition of the adsorbent after CeO_2 doping is changed on the microscopic scale and also to reveal the interaction mechanism between Hg^0 and the AC surface, especially when only experimental evidence can be relied on. In addition, during the process of Hg^0 removal, there may be complex processes such as chemical adsorption and oxidation of Hg^0 catalyzed by activated carbon, yet the mechanism of these processes has not been clearly unveiled. To date, few theoretical researches on Hg^0 removal by CeO_2 modified AC surface have been reported. Therefore, DFT calculations were conducted in this research to study the Hg^0 removal mechanism. A single-layer, zigzag graphene model was constructed to simulate the basic structure of AC adsorbents. The adsorption mechanism of mercury species was proposed from the analyses of stable geometries, adsorption energies, and electronic structures. Moreover, the oxidation pathway, energy barriers, and transition states were also obtained to investigate the Hg^0 oxidation mechanism. To better clarify the promotion effect of CeO_2 doping, a series of calculations was also performed on the plain AC surface as a comparison.

2. Models and Computational Details

2.1. AC and CeO2-AC Models

Experimental results have indicated that the surface structure of carbon-based adsorbents is mainly composed of 3–7 randomly associated benzene rings [24,25]. In terms of quantum chemical calculations, it has been reported that a carbonaceous species such as AC is generally represented by a single layer of graphene. The unsaturated edge can be the activated sites for adsorption because these carbon atoms have unpaired electrons that are easy to transfer [26,27]. Therefore, a zigzag-shaped graphene model which contained 6-fused benzene rings ($C_{24}H_8$) was applied in this paper as the basic framework. As for CeO_2 modification, a CeO_2 molecule was adsorbed on the middle edge of graphene to simulate the CeO_2-AC surface. The sorbent surface model after geometry optimization is shown in Figures 1 and 2. Relevant geometric parameters are listed in Table 1.

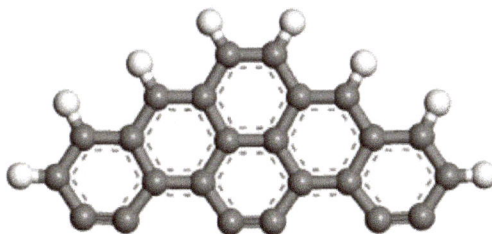

Figure 1. Conventional activated carbon (AC) surface model (the gray spheres represent C atoms, the white spheres represent H atoms, the same below).

Figure 2. CeO_2-AC surface model (red spheres represent O atoms, light yellow spheres represent Ce atoms, the same below).

Table 1. Geometric parameters of activated carbon (AC) surface model.

Bond Type		C–C	C–H	C–O	C≡C	Ce–O
Bond Length (Å)	Conventional AC model	1.369–1.439	1.092	-	1.267	-
	CeO_2 modified AC model	1.394–1.460	1.093	1.348	1.275	2.217

2.2. Computational Parameters

All DFT calculations were conducted by the DMol^3 program in this research [28]. The GGA-PBE method was utilized to describe the exchange-correlation functional [29,30]. The core DFT semi-core pseudopotential (DSPP) method was used to set the core treatment of Hg and Ce, for C, H, Cl and O, the all-electron method was applied [31]. A real space orbital cutoff of 4.5 Å was conducted and a 0.005 Ha smearing was used to facilitate the self-consistent field (SCF) convergence. The double numerical basis set plus polarization (DNP method) with p-functions on all hydrogen atoms was applied for molecular orbitals in this research. The thresholds of energy, force, and displacement are 10^{-5} Hartree for energy change, 2×10^{-3} Ha/atom for the maximum force, and 5×10^{-3} Å for displacement. The adsorption energy on the AC surface (E_{ads}) is defined by Formula (1):

$$E_{ads} = E_{adsorbate-substrate} - E_{adsorbate} - E_{substrate} \qquad (1)$$

where $E_{adsorbate-substrate}$ denotes the energy of adsorption configurations on the AC surface, $E_{adsorbate}$ and $E_{substrate}$ denote the energies of adsorbate and AC model, respectively. According to Formula (1), a negative value of E_{ads} indicates an exothermic reaction and high negative value indicates more heat release and a more stable product.

The reaction pathway is composed of intermediate (IM), transition state (TS), and final state (FS). The complete linear synchronous transit and quadratic synchronous transit (LST/QST) method was applied to obtain all transition states in this study [32]. Vibrational frequencies were also calculated for optimized configurations to identify stationary points (no imaginary frequency) and transition states (only one imaginary frequency). Reaction barriers were obtained by Formula (2):

$$E_{barrier} = E_{transition\ state} - E_{intermediate} \qquad (2)$$

where $E_{transition\ state}$ and $E_{intermediate}$ represent the energies of transition states and intermediates, respectively.

3. Results and Discussion

3.1. Hg^0 Adsorption Mechanism

The Hg^0 adsorption mechanism was studied on the AC model and CeO_2-AC model first of all. All possible sites where the adsorption may occur were considered on the unsaturated edge, two stable structures (1A, 1B and 1A*, 1B*, shown in Figure 3) were obtained on the CeO$_2$-AC and AC surface, respectively. The adsorption energies and related geometry parameters of 1A, 1B, 1A* and 1B* are shown in Table 2. In Figure 3 and Table 2, in terms of CeO$_2$-AC surface, the distance between the Ce and Hg atom is 3.431 Å in 1A and 3.508 Å in 1B. The adsorption energy is in the order of 1B < 1A (−42.05 kJ/mol vs. −46.94 kJ/mol). Generally, the adsorption energy of physisorption is usually about 0.1 eV (9.65 kJ/mol), while the adsorption energy of chemisorption usually ranges from 2 to 3 eV (192.97–289.46 kJ/mol) for molecular adsorption on solid surfaces [33]. Hence, the adsorption energies of Hg^0 indicate a weak chemical adsorption. On the surface of AC, however, Hg^0 adsorption energy in 1A* and 1B* is −119.92 kJ/mol and −146.01 kJ/mol, respectively. The distance between the Hg^0 and AC surface is 2.511 Å for 1A* and 2.462/2.464 Å for the 1B* configuration. For the CeO$_2$-AC model, the Hg^0 adsorption energy is lower than that on the AC surface. Therefore, it can be concluded that the doping of the CeO$_2$ molecule weakens the binding of Hg^0 to some extent on the AC surface, but the adsorption manner of Hg^0 still belongs to chemical adsorption.

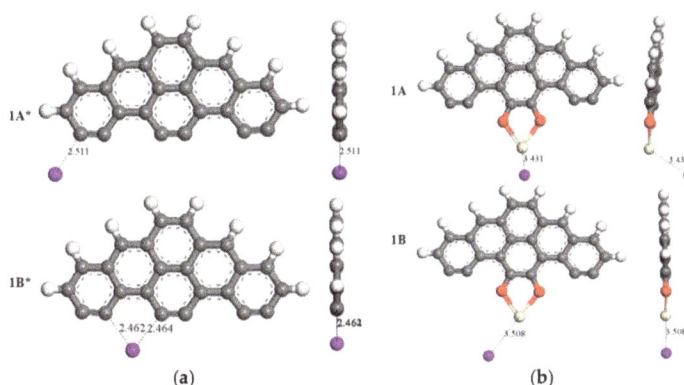

Figure 3. (a) Hg^0 adsorption configurations on CeO$_2$-AC surface; (b) Hg^0 adsorption configurations on AC surface. (The purple spheres represent Hg atoms, the same below. Both front view and side view are given.)

Table 2. Adsorption energies and configurations of Hg^0 on AC surface.

	CeO$_2$-AC Model			AC Model	
Configurations	R_{Hg-Ce} (Å)	E_{ads} (kJ/mol)	Configurations	R_{Hg-C} (Å)	E_{ads} (kJ/mol)
1A	3.431	−46.94	1A*	2.511	−119.92
1B	3.508	−42.05	1B*	2.462/2.464	−146.01

To further study the interaction between Hg^0 and the AC surface during adsorption, the partial density of states (PDOS) analysis for Hg, Ce and C atoms were conducted for 1A and 1B*, which are the most stable configurations. The PDOS diagrams for Hg pre/post-adsorption, Ce pre/post-adsorption, and C pre/post-adsorption are depicted in Figure 4. In Figure 4a,b, the s and p orbital peaks of Hg are shifted to lower energy level after adsorption because of the electron transfer between Hg^0 and the CeO$_2$-AC surface. At the same time, the intensity of the p-orbital is reduced, the peak of the Hg d-orbital at 16 eV disappears, and the peak at −3.1 eV also migrates to a lower energy level, which indicates that there is an intensive interplay between Hg^0 and the CeO$_2$-AC surface. For Ce atoms, the energy band before Hg^0 adsorption locates between −37 eV and 1.5 eV, all orbital peaks of the C atom on the AC surface only slightly change after adsorption while the peak of the d-orbit at 0.6 eV is slightly strengthened. The above analysis shows that Hg^0 is strongly adsorbed and interacts with the Ce atoms on the surface, the electron orbitals of the Hg atom change significantly, while the electronic structure of the Ce atoms on the CeO_2-AC surface does not change much, indicating the CeO_2-AC sorbent remained stable after adsorption.

The PDOS of C and Hg on the AC surface before and after adsorption are shown in Figure 4c,d. Similar to that on the CeO_2-AC surface, the s-orbital and p-orbital peaks of Hg after adsorption shift to lower energy level; the p-orbital intensity is reduced and retains several short peaks between -1.9 eV and 7.1 eV. The Hg d-orbital peak at 16 eV vanishes, the peak at -3.1 eV migrates to lower energy level, indicating an intensive interplay between Hg^0 and the CeO_2-AC surface. The intensity of the p-orbital peak of the C atom on the surface slightly weakens, while the s-orbital peak does not show any significant change. According to the above analysis, it is known that the PDOS of Hg does not change much after CeO_2 doping, both adsorption sites on Ce and C atoms remain stable after Hg^0 is strongly bound to the surface. Hence, it can be considered that although the doping of CeO_2 reduces the adsorption strength of Hg^0 on AC sorbents to some extent, the adsorption mechanism remains the same and the capture of Hg^0 in flue gas by AC sorbents is not significantly inhibited.

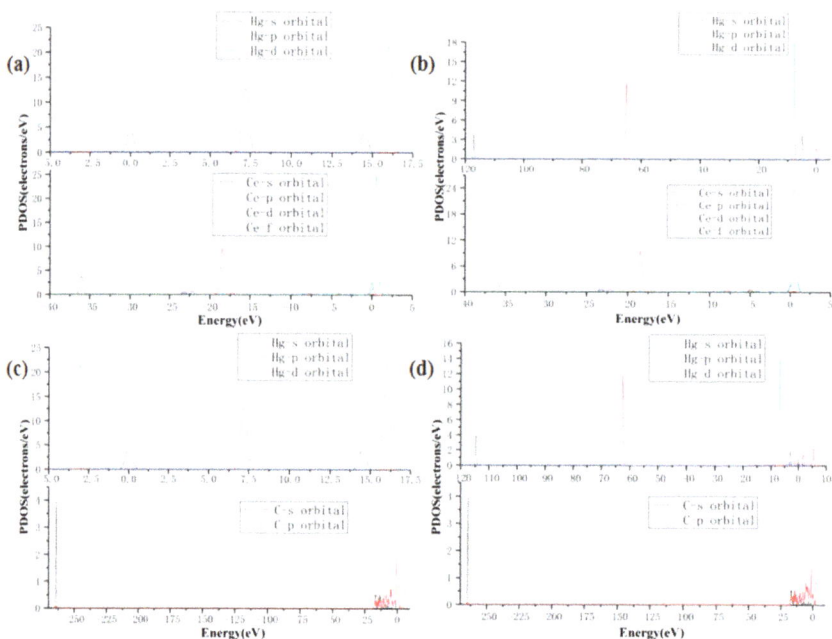

Figure 4. (a) Hg^0 pre-adsorption partial density of states (PDOS) on CeO_2-AC; (b) Hg^0 post-adsorption PDOS on CeO_2-AC; (c) Hg^0 pre-adsorption PDOS on AC; (d) Hg^0 post-adsorption PDOS on AC.

3.2. HCl Adsorption Mechanism

Previous studies have shown that HCl is a crucial compound in the Hg^0 oxidation process, because the Hg species in flue gas mainly exists as $HgCl_2$ after Hg^0 oxidation [34]. Therefore, it is necessary to explore the adsorption manner of HCl, $HgCl_2$ and intermediate HgCl to reveal the oxidation mechanism of Hg^0 in flue gas. DFT calculations were applied to investigate all possible adsorption configurations of HCl molecules on the surface of CeO_2-AC and AC. A stable configuration and adsorption energy were obtained and are shown in Figure 5 and Table 3.

As is seen in Figure 5, the adsorption mode of HCl molecules can be divided into two categories. (1) The HCl molecule dissociates on the AC surface, as shown by configurations 2A–2B, 2A*–2C*, which have relatively high adsorption energies (-149.38 kJ/mol and -371.63 kJ/mol), the binding of HCl molecules to the surface is relatively intense. (2) No dissociation occurs on the sorbents, as depicted in configuration 2C–2E and 2D*–2E*, the adsorption energies range from −41.20 kJ/mol to −55.79 kJ/mol, suggesting a weak chemical adsorption. In these configurations, HCl exists on the surface as HCl molecules, the geometric parameters and adsorption energies are close, only the orientation of the HCl molecule and adsorption site are slightly different. By comparing the adsorption configuration and adsorption energy of HCl on the two AC surfaces, it can be concluded that after CeO_2 doping, the adsorption energy of the HCl molecules increases, and the adsorption capacity for HCl is enhanced.

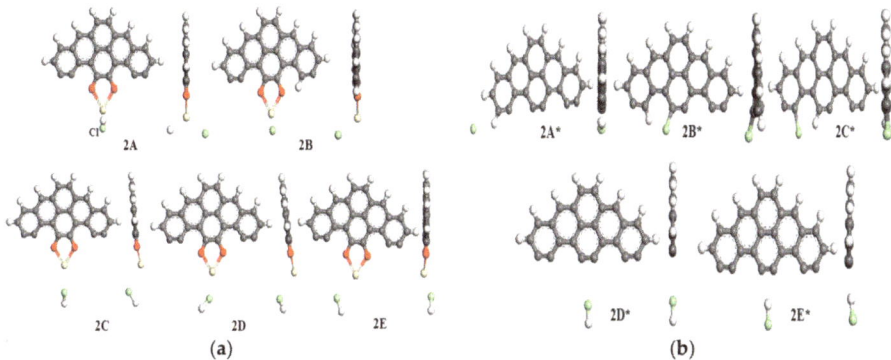

Figure 5. (**a**) Adsorption configurations of HCl on CeO_2-AC model; (**b**) Adsorption configurations of HCl on AC model. (Green spheres represent Cl atoms, the same below. Front view and side view of configurations are all given.)

Table 3. Adsorption energies/configurations of HCl on CeO_2-AC surface and AC surface.

	CeO₂-AC Surface				AC Surface		
Configurations	R_{X-Ce}(Å) *	R_{H-Cl}(Å)	E_{ads}(kJ/mol)	Configurations	R_{X-C}(Å) *	R_{H-Cl}(Å)	E_{ads}(kJ/mol)
2A	2.633/2.157	4.036	−174.27	2A*	1.095	2.707	−149.38
2B	2.623	5.291	−371.63	2B*	2.373	1.089	−254.78
2C	3.088	1.361	−54.85	2C*	2.257	1.088	−255.40
2D	3.071	1.362	−54.97	2D*	3.496/3.501	1.291	−41.20
2E	3.226	1.343	−52.63	2E*	2.411/2.385	1.304	−55.79

* X denotes H or Cl atoms.

3.3. HgCl Adsorption Mechanism

HgCl is an important intermediate for Hg^0 oxidation and can be further oxidized to $HgCl_2$. The stable adsorption configuration and adsorption energy of HgCl on the active site of AC sorbent are shown in Figures 6 and 7 and Table 4. The adsorption manner of HgCl on the AC surface can also be divided into two kinds, dissociative adsorption and non-dissociative adsorption. In configuration 3A and 3B, dissociation occurs for the HCl molecule and the Hg–Cl bond is cracked. The distance between the Cl radical and the Ce atom on the adsorbent surface is 2.623 Å/2.625 Å and the adsorption energies of the two configurations are −281.58 kJ/mol and −294.50 kJ/mol, respectively. In terms of 3C and 3D, the Hg–Cl bond is not broken, the bond lengths are 2.556 Å, 2.580 Å, and the adsorption energy of HgCl molecules on the AC surface are −153.09 kJ/mol and −139.80 kJ/mol, respectively. Both kinds of adsorption belong to intense chemical adsorption, suggesting that the combination of HgCl on the sorbents is relatively stable, which is beneficial to the conversion of the intermediate HgCl to the final product $HgCl_2$.

The adsorption mechanism of HgCl on AC is similar to that on the CeO_2-AC surface. In the 3A* and 3B* configurations, the Hg atom of HgCl is bound to a C atom on the AC surface, the adsorption energies are −258.70 kJ/mol and −259.47 kJ/mol, respectively. In the 3C* configuration, HgCl molecules are cracked, the Hg atom and Cl radical are separately combined with two carbon atoms on the graphene edge, the adsorption energy is -301.32 kJ/mol. It can be seen that the doping of CeO_2 lowers the adsorption energy of HgCl slightly, while the adsorption mechanism remains the same.

Table 4. Adsorption energies/configurations of HCl on CeO_2-AC model and AC model.

CeO₂-AC Surface				AC Surface			
Configurations	R_{X-Ce} (Å) *	R_{Hg-Cl} (Å)	E_{ads} (kJ/mol)	Configurations	R_{Hg-C} (Å) *	R_{Hg-Cl} (Å)	E_{ads} (kJ/mol)
3A	2.623	7.191	-281.58	3A*	2.149	2.391	-258.70
3B	2.625	5.219	-294.50	3B*	2.156	2.393	-259.47
3C	3.177	2.556	-153.09	3C*	2.323/2.427	3.689	-301.32
3D	3.275	2.580	-139.80	-	-	-	-

* X denotes Hg or Cl atoms.

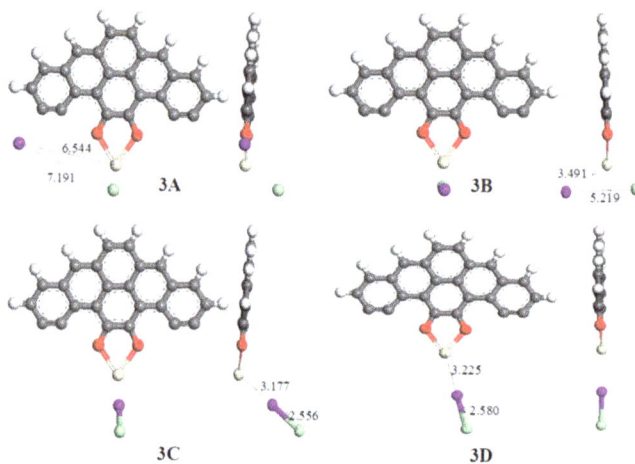

Figure 6. Adsorption configurations of HgCl on CeO_2-AC model.

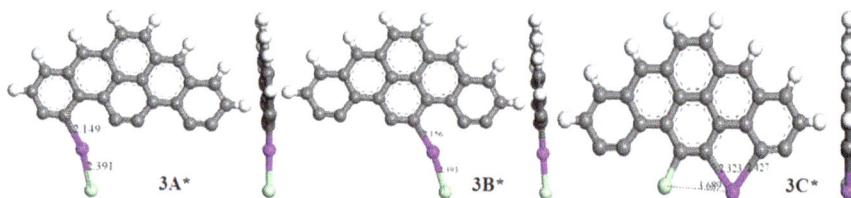

Figure 7. Adsorption configurations of HgCl on AC model.

3.4. HgCl₂ Adsorption Mechanism

When HCl exists in flue gas, $HgCl_2$ is the final product from Hg^0 oxidation. All possible sites on the edge of graphene have been considered, configurations and adsorption energies are depicted in Figures 8 and 9 and Table 5. For 4A configuration on the CeO_2-AC surface, $HgCl_2$ dissociated into an HgCl radical and a Cl radical, which is bound to the CeO_2-AC surface by chemisorption, the adsorption energy for this process is -202.70 kJ/mol. For configuration 4B, the $HgCl_2$ molecule also dissociates into an HgCl radical and a Cl radical, which are all bound to the CeO_2-AC surface with -287.46 kJ/mol adsorption energy. In configuration 4C, two Hg–Cl bonds of the $HgCl_2$ molecule are all broken, forming one weakly adsorbed Hg atom on the edge of the CeO_2-AC and two Cl radicals bonded to the Ce atom supported on the CeO_2-AC surface. Since $HgCl_2$ is completely dissociated, the adsorption energy of the configuration 4C is the largest, i.e., -399.03 kJ/mol. Based on the above results, after the $HgCl_2$ molecule is captured by the CeO_2-AC sorbent, it may exist on the AC sorbent in the manner of dissociative adsorption, and the value of the adsorption energies suggests that the bonding is considerably strong. Moreover, it is noteworthy that the adsorption energy in 4C is the largest, and the adsorption energy of the 4B configuration is the second largest when the Cl radical and Hg atoms are adsorbed on the surface, when only one Cl radical is bonded in 4A, is the adsorption energy minimal. Hence, it can be inferred that the adsorption energy of Hg^0 is smaller than that of the Cl radical on the surface; the adsorption can be stronger when the Cl end of the $HgCl_2$ molecule is bonded on the adsorbent.

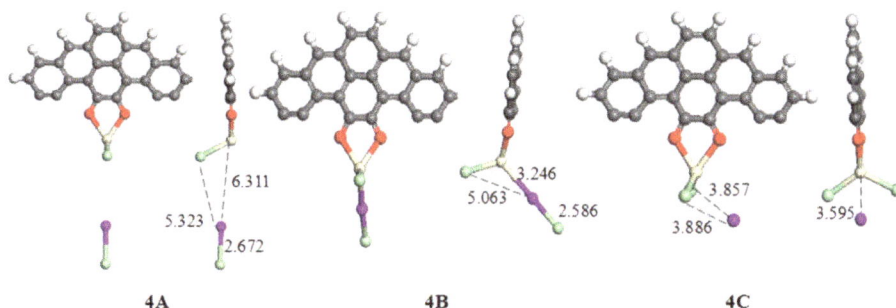

Figure 8. Adsorption configurations of $HgCl_2$ on CeO_2-AC model.

On the surface of the plain AC surface without CeO_2, however, the adsorption of $HgCl_2$ includes two types, dissociative adsorption and non-dissociative adsorption. In terms of 4A*, one Hg–Cl bond of the $HgCl_2$ molecule is broken, the generated Cl radical is bound to the AC surface and the Hg end of another part of the HgCl radical is also bound to the adjacent carbon atom on the plain AC surface, the adsorption energy of 4A* is −369.19 kJ/mol. In configuration 4B* where non-dissociative adsorption occurs, the Hg–Cl bonds are slightly elongated but not broken compared to that of the $HgCl_2$ molecule (2.345 Å) in the gas phase, and the linear molecule configuration remains steady. The adsorption energy on the plain AC surface is −101.54 kJ/mol. In the 4C* configuration, the bond angle ∠Cl–Hg–Cl is slightly reduced compared with the $HgCl_2$ molecule in the gas phase (180°), the Hg–Cl bond length is slightly elongated, the $HgCl_2$ molecule is adsorbed on the AC model in parallel, the adsorption energy is −110.80 kJ/mol. For configuration 4D* the $HgCl_2$ molecule is distorted, the Hg–Cl bond elongates slightly and the angle of the ∠Cl–Hg–Cl is reduced to 92.65° with a −186.98 kJ/mol adsorption energy. Compared with the dissociative adsorption configurations on the CeO_2-AC surface, the adsorption energy on the conventional AC surface is more positive, indicating that the $HgCl_2$ molecule is preferred to be dissociatively adsorbed on the AC after CeO_2 doping, the $HgCl_2$ molecule is more tightly bound on the CeO_2-AC surface, so it has a better removal performance for $HgCl_2$ in coal-fired flue gas.

Table 5. Adsorption energies and configurations of $HgCl_2$ on CeO_2-AC surface and AC surface.

	CeO$_2$-AC Surface				AC Surface			
Configurations	R_{X-Ce} (Å)	R_{Hg-Cl} (Å)	E_{ads} (kJ/mol)	Configurations	R_{Hg-C} (Å)	R_{Hg-Cl} (Å)	$\angle$Cl–Hg–Cl (°)	E_{ads} (kJ/mol)
4A	2.599	5.323/2.672	−202.70	4A*	2.121	2.368/3.468	-	−369.19
4B	2.616	5.063/2.586	−287.46	4B*	6.043/6.119	2.356/2.362	179.82	−101.54
4C	2.640/2.643	3.886/3.857	−399.03	4C*	3.202/3.258	2.367/2.368	171.63	−110.80
-	-	-	-	4D*	2.271/2.309	2.546/2.515	92.65	−186.98

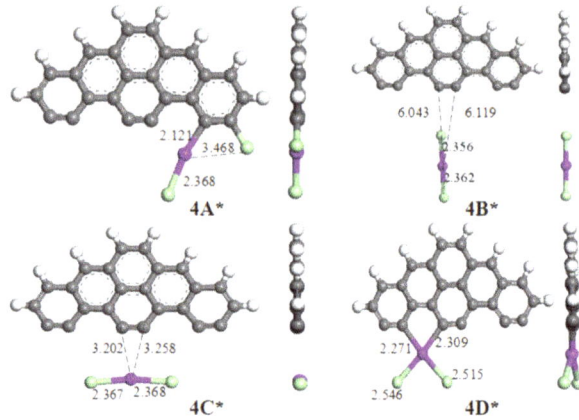

Figure 9. Adsorption configurations of HgCl$_2$ on AC model.

3.5. Hg^0 Oxidation Mechanism

Based on the Hg^0, HCl, HgCl, and HgCl$_2$ adsorption mechanism discussed above, the oxidation mechanisms of Hg^0 on the CeO$_2$-AC and AC surface were further studied, the reaction energy barriers and relative geometries in the oxidation route were calculated. The potential energy diagram of the reaction pathways are illustrated in Figure 10, and the relative configurations for reactants, products, and transition states involved in the paths are shown in Figure 11.

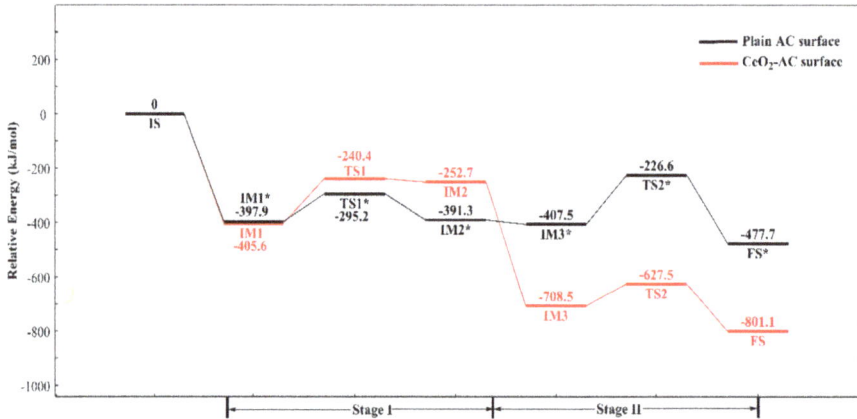

Figure 10. Potential energy diagram of Hg^0 oxidation pathways on CeO$_2$-AC surface and plain AC surface.

The red line in Figure 10 represents the oxidation pathway of Hg^0 on the CeO$_2$-AC surface. Depending on the different products, the oxidation path can be divided into two phases. In the first phase, the HCl molecule is first combined with the sorbent surface by dissociative adsorption. Then an H radical and a Cl radical are adsorbed on the Ce atom on the CeO$_2$-AC surface, releasing 405.6 kJ/mol to form a stable IM1 configuration. Subsequently, the Hg^0 atom and Cl radical migrate toward each other forming the transition state TS1, a 2.689 Å Hg–Cl bond is formed and finally stabilized at 2.572 Å in IM2. In this process, an HgCl molecule is formed with a reaction barrier of 165.3 kJ/mol and 152.9 kJ/mol adsorption energy. After the HgCl is formed, the oxidation reaction in the second stage is carried out according to the path of IM2 → IM3 → TS2 → FS. Firstly, the IM2 configuration combines with another HCl molecule to form a stable configuration, IM3, which is exothermic by 450.8 kJ/mol. In IM3, another adsorbed HCl molecule is cracked. After dissociation, the Hg–Cl bond is formed between the Cl radical and HgCl to form a bent $HgCl_2$ molecule, while the H radical moves close to the C atom at the edge of AC surface, forming the transition state TS2, with an 81.0 kJ/mol energy barrier for this process. Consequently, in the transition state TS2, a linear $HgCl_2$ molecule further desorbs from the CeO$_2$-AC surface to form the final state FS. The distance between the Hg and Ce atom on the AC surface in FS is expanded from 3.212 Å to 6.470 Å; the process generating $HgCl_2$ requires an exotherm of 92.5 kJ/mol. It can be seen that the two-step oxidation of Hg^0 and HCl follows the Langmuir–Hinshelwood mechanism [35,36], and the rate-determining step is the first stage, i.e., the formation of the HgCl molecule. The energy barrier of the entire pathway is 165.3 kJ/mol.

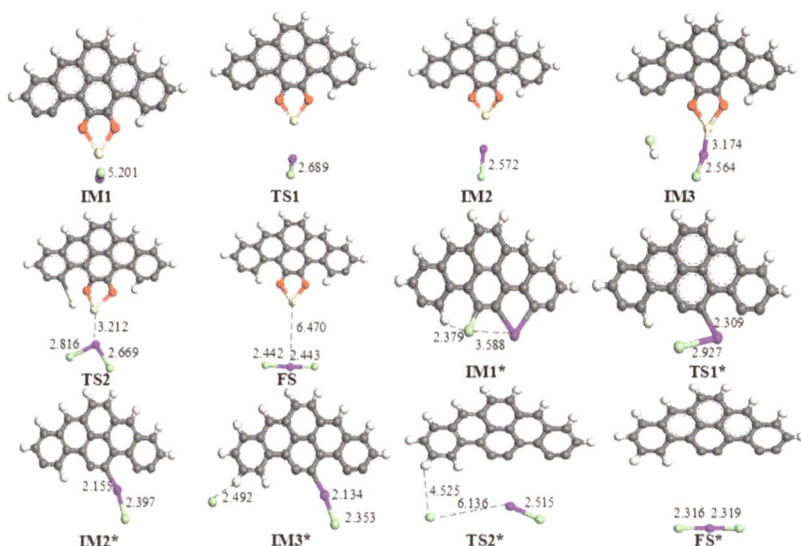

Figure 11. Relative configurations of reactions, products, and intermediate states involved in Hg^0 oxidation route.

The oxidation pathway of Hg^0 on the AC surface is depicted as the black portion in Figure 10, and the entire reaction is still divided into two phases. The first phase is the formation process of HgCl. First of all, one Hg atom and one HCl molecule are adsorbed on the AC surface to form IM1* and release 397.9 kJ/mol of heat, the HCl molecule dissociates into one Cl radical and one H radical respectively combining with two C atoms on the AC surface. Subsequently, the Cl radical migrates to the Hg atom adsorbed on the adjacent C site, forming a transition state TS1* and overcoming a 102.7 kJ/mol barrier. In TS1*, the Cl radical further moves closely to Hg^0, the distance between the Cl radical and Hg^0 in IM2* is shortened from 2.927 Å to 2.397 Å, the obtained HgCl molecule is continuously adsorbed on the AC surface. In the second phase, the IM2* configuration captures another HCl molecule and releases 16.2 kJ/mol of heat converting to the configuration IM3*, the Hg–Cl bond length in IM3* is 2.353 Å. Subsequently, HCl in the transition state TS2* dissociates again on the AC surface, the distance between the H radical and Cl radical increases from 2.492 Å to 4.525 Å, the distance between the Cl radical and Hg atoms is 6.136 Å, and the energy barrier to form transition state TS2* is 180.9 kJ/mol. In configuration TS2*, the dissociated Cl radical continues to migrate to the HgCl molecule, eventually forming $HgCl_2$ in FS* which is exothermic by 70.2 kJ/mol to form $HgCl_2$. The reaction between the Hg^0 and HCl molecule in the conventional AC model also follows the Langmuir–Hinshelwood mechanism. However, due to the high energy barrier in the second phase of oxidation, unlike the CeO$_2$-AC surface, the rate-determining step is the formation of $HgCl_2$.

Based on the above discussion, the adsorption energy of Hg^0 and HgCl on the AC sorbents is reduced after CeO_2 modification, the interplay between Hg^0 and the AC sorbent is relatively weak, resulting in a relatively high energy barrier in the formation of HgCl in the first stage of Hg^0 oxidation. On the other hand, the modification of CeO_2 increases the adsorption energy of HCl and $HgCl_2$ on the AC sorbent, which is beneficial to the further oxidation of the intermediate HgCl to form $HgCl_2$, making the reaction energy barrier relatively low. Therefore, on the CeO_2-AC surface, the energy barrier of the HgCl formation process is relatively high, which is the rate-determining step of Hg^0 oxidation, whereas, on the plain AC model, the rate-determining step is the $HgCl_2$ formation process. Comparing the energy barriers on these two AC sorbents, the modification of CeO_2 reduces the reaction energy barrier of Hg^0 oxidation to some extent (165.3 kJ/mol vs. 180.9 kJ/mol). Ultimately, the catalytic oxidation capacity of AC sorbents is enhanced for removing mercury species in coal-fired flue gas.

Moreover, adsorption energies of HCl and $HgCl_2$ demonstrate that the adsorption of HCl and $HgCl_2$ on CeO_2 is stronger than that on the C atoms on the plain AC surface. Therefore, although reducing the adsorption strength of Hg^0 on the AC surface to some extent, the doped CeO_2 forms some adsorption sites which are easier for HCl and $HgCl_2$ to be adsorbed. In addition, for heterogeneous oxidation of mercury, the route involves two steps, i.e., the formation of HgCl and $HgCl_2$, and the latter is usually regarded as the rate-determining step due to the relatively high energy barrier. Whereas, with regard to the CeO_2-AC sorbents, the energy barrier of $HgCl_2$ formation is reduced on the CeO_2 site, hence promoting the oxidation procedure of Hg^0 and improving the ability to capture mercury species.

4. Conclusions

DFT calculations were implemented in this paper to study the Hg^0 adsorption and oxidation mechanism in coal-fired flue gas on a plain AC surface and CeO_2-AC surface. The adsorption energies were calculated, as well as the energy barriers of the Hg^0 oxidation pathways. The effect of CeO_2 doping was estimated by comparison of the removal mechanism on the plain AC surface and CeO_2-AC surface.

The calculation results demonstrate that the adsorption modes of Hg^0, HCl, HgCl and $HgCl_2$ are chemisorption, indicating they are tightly bound on the AC sorbent. PDOS analysis of the Hg^0 adsorption shows that Hg^0 is intensely adsorbed on the adsorbent, strong interaction and charge transfer occur between Hg^0 and the AC surface, the orbital peaks in the PDOS figure change significantly after adsorption. HCl, HgCl and $HgCl_2$ can all be dissociated and adsorbed on the AC surface, after dissociation of HCl molecules, the Cl radical does not bond with the surface. Hg and HgCl in the gas phase may combine with a Cl radical and generate the intermediate HgCl or the final product $HgCl_2$. The adsorption energies of HgCl and $HgCl_2$ are relatively high, indicating that CeO_2 doping enhances the ability to capture HgCl and $HgCl_2$. What is more, the AC sorbent has a certain catalytic effect on the oxidation of Hg^0, the rate-determining step of Hg^0 oxidation on the plain AC surface is the formation of $HgCl_2$. The doping of CeO_2 reduces the reaction energy barrier and enhances the catalytic oxidation ability to remove Hg^0. The rate-determining step is the generation of HgCl. On the plain AC surface and CeO_2-AC surface, the oxidation of Hg^0 follows the Langmuir–Hinshelwood mechanism.

Author Contributions: Conceptualization, L.Z.; Data curation, Y.-w.W., J.H. and H.-x.W.; Formal analysis, Y.-w.W., J.H., H.-x.W., D.-j.L., Q.L. and Y.-p.Y.; Funding acquisition, Q.L.; Project administration, Q.L.; Writing—original draft, L.Z.; Writing—review and editing, L.Z., Y.-w.W., J.H., H.-x.W., D.-j.L., Q.L. and Y.-p.Y.

Funding: This research was funded by the National Basic Research Program of China (2015CB251501), National Natural Science Foundation of China (51876060), Beijing Nova Program (Z171100001117064), Beijing Natural Science Foundation (3172030), Grants from Fok Ying Tung Education Foundation (161051), and Fundamental Research Funds for the Central Universities (2018ZD08, 2016YQ05).

Conflicts of Interest: The authors declare no conflict of interest.

References

1. Wang, S.X.; Zhang, L.; Wang, L.; Wu, Q.R.; Wang, F.Y.; Hao, J.M. A review of atmospheric mercury emissions, pollution and control in China. *Front. Environ. Sci. Eng.* **2014**, *8*, 631–649. [CrossRef]
2. Streets, D.G.; Lu, Z.F.; Levin, L.; Ter Schure, A.; Sunderland, E.M. Historical releases of mercury to air, land, and water from coal combustion. *Sci. Total Environ.* **2017**, *615*, 131–140. [CrossRef] [PubMed]
3. Ji, W.C.; Shen, Z.M.; Tang, Q.L.; Yang, B.W.; Fan, M.H. A DFT study of Hg^0, adsorption on Co_3O_4 (110) surface. *Chem. Eng. J.* **2016**, *289*, 349–355. [CrossRef]
4. Liu, Z.J.; Zhang, Z.Y.; Choi, S.K.; Liu, Y.Y. Surface properties and pore structure of anthracite, bituminous coal and lignite. *Energies* **2018**, *11*, 1502. [CrossRef]
5. Wu, Y.; Wang, S.X.; Streets, D.G.; Hao, J.M.; Chan, M.; Jiang, J.K. Trends in anthropogenic mercury emissions in China from 1995 to 2003. *Environ. Sci. Technol.* **2006**, *40*, 5312–5318. [CrossRef] [PubMed]
6. Li, H.L.; Zhu, L.; Wu, S.K.; Liu, Y.; Shih, K.M. Synergy of CuO and CeO_2 combination for mercury oxidation under low-temperature selective catalytic reduction atmosphere. *Int. J. Coal Geol.* **2017**, *170*, 69–76. [CrossRef]
7. Zhao, L.; Liu, Y.; Wu, Y.W.; Han, J.; Zhang, S.L.; Lu, Q.; Yang, Y.P. Mechanism of heterogeneous mercury oxidation by HCl on V_2O_5(001) surface. *Curr. Appl. Phys.* **2018**, *18*, 626–632. [CrossRef]
8. Zhao, L.; He, Q.S.; Li, L.; Lu, Q.; Dong, C.Q.; Yang, Y.P. Research on the catalytic oxidation of Hg^0 by modified SCR catalysts. *J. Fuel Chem. Technol.* **2015**, *43*, 628–634. [CrossRef]
9. Zhou, Z.J.; Liu, X.W.; Zhao, B.; Chen, Z.G.; Shao, H.Z.; Wang, L.L.; Xu, M.H. Effects of existing energy saving and air pollution control devices on mercury removal in coal-fired power plants. *Fuel Process. Technol.* **2015**, *131*, 99–108. [CrossRef]
10. Huang, W.J.; Xu, H.M.; Qu, Z.; Zhao, S.; Chen, W.; Yan, N. Significance of Fe_2O_3, modified SCR catalyst for gas-phase elemental mercury oxidation in coal-fired flue gas. *Fuel Process. Technol.* **2016**, *149*, 23–28. [CrossRef]
11. Fernández-Miranda, N.; Rodriguez, E.; Lopez-Anton, M.A.; Garcia, R.; Martinez-Tarazona, M.R. A New approach for retaining mercury in energy generation processes: Regenerable carbonaceous sorbents. *Energies* **2017**, *10*, 1311. [CrossRef]
12. Gao, X.P.; Zhou, Y.N.; Tan, Y.J.; Cheng, Z.W.; Tang, Z.W.; Jia, J.P. Unveiling adsorption mechanisms of elemental mercury on defective boron nitride monolayer: A computational study. *Energy Fuels* **2018**, *32*, 5331–5337. [CrossRef]
13. Zhao, L.; Wu, Y.; Han, J.; Lu, Q.; Yang, Y.; Zhang, L. Mechanism of mercury adsorption and oxidation by oxygen over the CeO_2 (111) surface: A DFT study. *Materials* **2018**, *11*, 485. [CrossRef] [PubMed]
14. Li, H.L.; Wu, C.Y.; Li, Y.; Zhang, J.Y. CeO_2-TiO_2 catalysts for catalytic oxidation of elemental mercury in low-rank coal combustion flue gas. *Environ. Sci. Technol.* **2011**, *45*, 7394–7400. [CrossRef] [PubMed]
15. Jampaiah, D.; Samuel, J.I.; Ylias, S.; James, T.; Selvakannan, P.R.; Ayman, N.; Benjaram, M.R.; Suresh, B. Ceria-zirconia modified MnOx catalysts for gaseous elemental mercury oxidation and adsorption. *Catal. Sci. Technol.* **2016**, *6*, 1792–1803. [CrossRef]
16. Jampaiah, D.; Katie, T.; Perala, V.; Samuel, J.I.; Ylias, S.; James, T.; Suresh, B.; Benjaram, M.R. Catalytic oxidation and adsorption of elemental mercury over nanostructured CeO_2–MnOx catalyst. *RSC Adv.* **2015**, *38*, 30331–30341. [CrossRef]
17. Hua, X.Y.; Zhou, J.S.; Li, Q.K.; Luo, Z.Y.; Cen, K.F. Gas-phase elemental mercury removal by CeO_2 impregnated activated coke. *Energy Fuels* **2010**, *24*, 5426–5431. [CrossRef]
18. Wang, Y.; Li, C.T.; Zhao, L.K.; Xie, Y.E.; Zhang, X.N.; Zeng, G.M.; Wu, H.Y.; Zhang, J. Study on the removal of elemental mercury from simulated flue gas by Fe_2O_3-CeO_2/AC at low temperature. *Environ. Sci. Polut. Res.* **2016**, *23*, 5099–5110. [CrossRef] [PubMed]
19. Zhu, Y.C.; Han, X.J.; Huang, Z.P.; Hou, Y.Q.; Guo, Y.P.; Wu, M.H. Superior activity of CeO_2 modified V_2O_5/AC catalyst for mercury removal at low temperature. *Chem. Eng. J.* **2017**, *331*, 741–749. [CrossRef]
20. Tian, L.H.; Li, C.T.; Li, Q.; Zeng, G.M.; Gao, Z.; Li, S.H.; Fan, X.P. Removal of elemental mercury by activated carbon impregnated with CeO_2. *Fuel* **2009**, *88*, 1687–1691. [CrossRef]
21. Xie, Y.E.; Li, C.T.; Zhao, L.K.; Zhang, J.; Zeng, G.M.; Zhang, X.N.; Zhang, W.; Tao, S.S. Experimental study on Hg^0 removal from flue gas over columnar MnOx-CeO_2/activated coke. *Appl. Surf. Sci.* **2015**, *333*, 59–67. [CrossRef]
22. Zhang, H.W.; Chen, J.Y.; Zhao, K.; Niu, Q.X.; Wang, L. Removal of vapor-phase elemental mercury from simulated syngas using semi-coke modified by Mn/Ce doping. *J. Fuel Chem. Technol.* **2016**, *4*, 394–400. [CrossRef]

23. Wu, H.Y.; Li, C.T.; Zhao, L.K.; Zhang, J.; Zeng, G.M.; Xie, Y.E.; Zhang, X.N.; Wang, Y. Removal of gaseous elemental mercury by cylindrical activated coke loaded with CoO$_x$-CeO$_2$ from simulated coal combustion flue gas. *Energy Fuels* **2015**, *29*, 6747–6757. [CrossRef]

24. Chen, N.; Yang, R.T. Ab initio molecular orbital calculation on graphite: Selection of molecular system and model chemistry. *Carbon* **1998**, *36*, 1061–1070. [CrossRef]

25. Yang, F.H.; Yang, R.T. Ab initio molecular orbital study of adsorption of atomic hydrogen on graphite: Insight into hydrogen storage in carbon nanotubes. *Carbon* **2002**, *40*, 437–444. [CrossRef]

26. Chen, P.; Gu, M.Y.; Lin, Y.Y.; Yan, D.W.; Huang, Y.Y. Effect of ketone group on heterogeneous reduction of NO by char. *J. Feul Chem. Technol.* **2018**, *46*, 521–528. [CrossRef]

27. Chompoonut, R.; Vinich, P.; Supa, H.; Nawee, K.; Supawadee, N. Complete reaction mechanisms of mercury oxidation on halogenated activated carbon. *J. Hazard. Mater.* **2016**, *310*, 253–260. [CrossRef]

28. Delley, B. From molecules to solids with the DMol^3 approach. *J. Chem. Phys.* **2000**, *113*, 7756–7764. [CrossRef]

29. Perdew, J.P.; Burke, K.; Ernzerhof, M. Generalized gradient approximation made simple. *Phys. Rec. Lett.* **1996**, *77*, 3865–3868. [CrossRef] [PubMed]

30. Perdew, J.P.; Burke, K.; Wang, Y. Generalized gradient approximation for the exchange-correlation hole of a many-electron system. *Phys. Rev. B.* **1996**, *54*, 16533–16539. [CrossRef]

31. Delley, B. Hardness conserving semilocal pseudopotentials. *Phys. Rev. B.* **2002**, *66*, 155125. [CrossRef]

32. Halgren, T.A.; Lipscomb, W.N. The synchronous-transit method for determining reaction pathways and locating molecular transition states. *Chem. Phys. Lett.* **1977**, *49*, 225–232. [CrossRef]

33. Orimo, S.; Zuttel, A.; Schlapbach, L.; Majer, G.; Fukunaga, T.; Fujii, H. Hydrogen interaction with carbon nanostructures: Current situation and future prospects. *J. Alloy. Compd.* **2003**, *356*, 716–719. [CrossRef]

34. Zhao, L.K.; Li, C.T.; Wang, Y.; Wu, H.Y.; Gao, L.; Zhang, J.; Zeng, G.M. Simultaneous removal of elemental mercury and NO from simulated flue gas using a CeO$_2$ modified V$_2$O$_5$-WO$_3$/TiO$_2$ catalyst. *Catal. Sci. Technol.* **2016**, *6*, 420–430. [CrossRef]

35. Presto, A.A.; Granite, E.J. Survey of catalysts for oxidation of mercury in flue gas. *Environ. Sci. Technol.* **2006**, *40*, 5601–5609. [CrossRef] [PubMed]

36. Zhang, B.K.; Liu, J.; Shen, F.H. Heterogeneous mercury oxidation by HCl over CeO$_2$ catalyst: Density functional theory study. *J. Phys. Chem. C* **2015**, *119*, 15047–15055. [CrossRef]

energies

MDPI

Article

Study on Powder Coke Combustion and Pollution Emission Characteristics of Fluidized Bed Boilers

Chen Yang [1,2,*], Haochuang Wu [1,2], Kangjie Deng [1,2], Hangxing He [3] and Li Sun [1,2]

1   Key Laboratory of Low-Grade Energy Utilization Technologies and Systems, Chongqing University, Ministry of Education, Chongqing 400030, China; haochuang@cqu.edu.cn (H.W.); kangjie@cqu.edu.cn (K.D.); sunli@cqu.edu.cn (L.S.)
2   School of Energy and Power Engineering, Chongqing University, Chongqing 400030, China
3   Science and Technology on Reactor System Design Technology Laboratory, Nuclear Power Institute of China, Chengdu 610213, China; hangxing@cqu.edu.cn
*   Correspondence: yxtyc@cqul.edu.cn

Received: 22 February 2019; Accepted: 9 April 2019; Published: 13 April 2019

Abstract: The fluidized reactor is widely used in a number of chemical processes due to its high gas-particle contacting efficiency and excellent performance on solid mixing. An improved numerical framework based on the multiphase particle-in-cell (MP-PIC) method has been developed to simulate the processes of gas–solid flow and chemical reactions in a fluidized bed. Experiments have been carried out with a 3-MW circulating fluidized bed with a height of 24.5 m and a cross section of 1 m^2. In order to obtain the relationship between pollutant discharge and operating conditions and to better guide the operation of the power plant, a series of tests and simulations were carried out. The distributions of temperature and gas concentration along the furnace from simulations achieved good accuracy compared with experimental data, indicating that this numerical framework is suitable for solving complex gas–solid flow and reactions in fluidized bed reactors. Through a series of experiments, the factors affecting the concentration of NO_x and SO_x emissions during the steady-state combustion of the normal temperature of powder coke were obtained, which provided some future guidance for the operation of a power plant burning the same kind of fuel.

Keywords: fluidized bed; powder coke; MP-PIC method; emission characteristics

1. Introduction

The fluidized reactor is widely used in a number of chemical processes (e.g., circulating fluidized bed boiler combustion, biomass pyrolysis, catalytic reaction, and conceptual fluidized bed nuclear reactor [1–4]) due to its high gas-particle contacting efficiency, wide fuel adaptability, and excellent performance on solid mixing. Powder coke, which is also called semicoke or coke powder, was used as fuel in this fluidized bed boiler research. The powder coke used in the experiment was a high-quality Jurassic coal block produced in China's Shenfu coalfield, which was carbonized and burned in an internally heated vertical retort. Its structure is blocky and its color is light black. It is a fuel and reducing agent for ferroalloys, fertilizers, calcium carbide, blast furnace injection, and other applications. As a new type of fluidized bed boiler fuel, it has a relatively high fixed carbon content; low ash, aluminum, and sulfur contents; and its specific resistance and chemical activity are high. With the slowdown of China's domestic thermal power market growth, higher requirements for energy savings and emissions reductions for new power plants are required. Thus, vigorous development of circulating fluidized bed (CFB) boiler combustion technology for special fuels has important strategic significance for broadening the future circulating fluidized bed market.

In order to study the fuel combustion and emission characteristics of fluidized bed boilers, a large number of experiments have been implemented over the past years to optimize the operating parameters of these fluidized bed boilers to achieve higher performance and lower costs. Further, extensive mathematical models of multiphase flow and reaction have been developed to explore the gas–solid flow reaction characteristics in fluidized bed boiler furnaces.

With the development of technology, the computing power of large computers continues to increase, and numerical methods have become a powerful tool to study the phenomenon of gas-solid fluidized flow reaction. Two theoretical approaches have been proposed to simulate the complex physical and chemical processes in a fluidized reactor, namely, the Eulerian-Eulerian method, also called the two-fluid model (TFM), and the hybrid Eulerian-Lagrangian model (HEL) [5–7]. The two-fluid model treats the solid phase as a pseudocontinuum; an analogy is made between the kinetic theory of granular flow with the molecular kinematic theory of dense gas. The macroscopic characteristics of fluidized reactors can be calculated with relatively small computational cost. This approach has been developed since it was first applied in the simulation of gas–solid flow in a riser by Sinclair and Jackson [8]. However, important parameters, such as particle size distribution and size variation due to reaction or collision, are very hard to describe with this method. Many unclosed terms are also generated during the solid particle averaging procedure, and methods to couple gas-phase turbulence with particle fluctuation still have their limits [9].

The hybrid Eulerian-Lagrangian approach, in which a particle is treated as a discrete element [10,11], is a good way to solve the gas–solid fluidized bed reaction. For the HEL model, each particle, composed of physical (size, temperature, and density) and chemical (reactive) properties, can be tracked individually, and detailed microscopic properties (particle trajectory and transient forces acting on a specific particle), which are extremely difficult to acquire by the TFM method, can also be obtained at the particle level. In this model, the computational cost increases rapidly when the number of particles calculated increases. Therefore, it is very difficult to achieve greater accuracy of the gas-particle flow field by simply adding the number of particles simulated. In order to overcome the contradiction of accuracy and computational cost, the multiphase particle-in-cell (MP-PIC) method is proposed to calculate the properties of particles in dense particle flow [12]. In the MP-PIC method, the solid phase is treated as both a continuum and a discrete phase, the particle stress gradient is calculated on the grid, and the particle properties are mapped from the Lagrangian coordinates to a Eulerian grid by the use of interpolation functions. This approach can attain a good equilibrium between CPU cost and the accuracy of results when simulating industrial-scale fluidized reactors.

Based on the MP-PIC method, Snider and Banerjee simulated a heterogeneous catalytic chemistry process in fluidized ozone decomposition and made a comparison between it with experimental results [13]. Abbsabi et al. simulated the 2D physical and chemical performance of the feeding section of a fast fluidized bed steam gasifier [14]. Xie et al. developed a 3D numerical model and used it to simulate coal gasification in a fluidized bed gasifier [11]. Ryan et al. used this method to investigate the performance of a carbon capture fluidized bed reactor [15]. Adamczyk et al. used this approach to model particle transport and combustion in an industrial-scale CFB boiler [16]. Loha et al. made a 3D kinetic simulation of fluidized bed biomass gasification by coupling the chemical reactions with the hydrodynamic calculation of gas–solid flow [17]. Zhong et al. modeled the olive cake combustion process in CFB with the gas flow solved by large eddy simulation [18]. In these studies, the calculated particles were treated as being of spherical shape, which is impossible in reality. Further, the variations of particles during the chemical process, such as fragmentation, shrink, and aggregation, were also neglected in these studies.

In this study, an improved numerical framework based on the MP-PIC method was developed to simulate the processes of gas–solid flow and chemical reactions in a fluidized bed. Experiments were carried out on a 3-MW circulating fluidized bed with a height of 24.5 m and a cross section of 1 m^2. Simulations and experiments were carried out under different operation conditions.

2. Mathematical Model/Numerical Method

2.1. Gas-Solid Hydrodynamics

2.1.1. Gas Phase

The gas-phase mass and momentum equations are as follows.
Gas-phase continuum equation:

$$\frac{\partial\left(\varepsilon_g\rho_g\right)}{\partial t}+\left(\nabla\cdot\varepsilon_g\rho_g\mathbf{u}_g\right)=\sum_{i=1}^{N_g}S_{gi} \tag{1}$$

where $\sum_{i=1}^{N_g}S_{gi}$ is the sum of the mass changes of each component due to the chemical reaction, and i is the number of gas-phase species.

Momentum equation:

$$\frac{\partial\left(\varepsilon_g\rho_g\mathbf{u}_g\right)}{\partial t}+\nabla\cdot\left(\varepsilon_g\rho_g\mathbf{u}_g\mathbf{u}_g\right)=-\varepsilon_g\nabla p+\nabla\cdot\left(\varepsilon_g\boldsymbol{\tau}_g\right)+\varepsilon_g\rho_g\mathbf{g}+\beta\left(\mathbf{u}_s-\mathbf{u}_g\right) \tag{2}$$

where β is the gas–solid drag coefficient [19].

$$\beta=\begin{cases}150\frac{\varepsilon_s^2\mu_g}{\varepsilon_g^2 d_s^2}+1.75\frac{\varepsilon_s\rho_g}{\varepsilon_g d_s}\left|\mathbf{u}_g-\mathbf{u}_s\right| & \varepsilon_g<0.8\\[2mm]\frac{3}{4}C_d\frac{\varepsilon_s\rho_g}{d_s}\left|\mathbf{u}_g-\mathbf{u}_s\right|\varepsilon_g^{-2.65} & \varepsilon_g\geq 0.8\end{cases} \tag{3}$$

$$C_d=\begin{cases}\frac{24}{Re}\left(1+0.15Re_s^{0.687}\right) & Re<1000\\[2mm]0.44 & Re\geq 1000\end{cases} \tag{4}$$

$$Re=\frac{2\rho_f\left|\mathbf{u}_g-\mathbf{u}_s\right|r_s}{\mu_g} \tag{5}$$

where μ_g is the molecular gas viscosity.

Energy equation:

$$\frac{\partial\left(\varepsilon_g\rho_g h_g\right)}{\partial t}+\nabla\cdot\left(\varepsilon_g\rho_g\mathbf{u}_g h_f\right)=$$
$$\varepsilon_g\left(\frac{\partial p}{\partial t}+\mathbf{u}_g\cdot\nabla p\right)+\Phi-\nabla\cdot\left[\varepsilon_g\left(-\lambda_g\nabla T_g\right)\right]+\dot{Q}+S_h+\dot{q}_D \tag{6}$$

where Φ represents the viscous dissipation term, $\dot{Q}$ represents the energy source term, and λ_g represents gas fluid thermal conductivity.

$\dot{q}_D$ represents enthalpy diffusion, written as

$$\dot{q}_D=\sum_{i=1}^{N_g}\nabla\cdot\left(h_i\varepsilon_g\rho_g D\nabla Y_{g,i}\right) \tag{7}$$

2.1.2. Solid Phase

Each component solid-phase particle was mapped to a discrete form by statistical weighting, and the m-th component particle having the same particle diameter D_m and density ρ_m was replaced by N_m calculated particles (i.e., parcels). For the M groups of fuel particles, the number of real particles N was calculated by; the spatial distribution of the parcels over time was $f\left\{\mathbf{X}_p^{(i)}(t),\mathbf{u}_p^{(i)}(t),D_p^{(i)},\rho_p^{(i)},W_p^{(i)},i=1,...,N\right\}$; and $\mathbf{X}_p^{(i)}(t),\mathbf{u}_p^{(i)}(t),D_p^{(i)},\rho_p^{(i)}$, and $W_p^{(i)}$ were the position, velocity, diameter, density, and statistical weight of a parcel. For the description of parcel motion, Newtonian mechanics was used for analysis. To simplify the discussion process, the superscript *i* indicating the *i*-th parcel was omitted.

The position equation of the parcel:

$$\frac{d\mathbf{X}_p(t)}{dt} = \mathbf{u}_p(t) \tag{8}$$

Equation of motion:

$$m_p \frac{d\mathbf{u}_p(t)}{dt} = m_p \mathbf{g} + \mathbf{F}_{p,\text{drag}}(t) + \mathbf{F}_{p,\text{coll}}(t) \tag{9}$$

where $\mathbf{g}$ is gravity, $\mathbf{F}_{p,\text{drag}}(t)$ is the drag force of the parcel (divided into solid-phase pressure gradient field drag force and gas–solid phase drag force), and $\mathbf{F}_{p,\text{coll}}(t)$ is the interaction force between the parcels.

Since the calculated particles use the same particle size as the actual particles, it was necessary to consider the interaction of W_p particles in the calculation of the drag force of the particles. For the *i*-th calculated particle of the *m*-th solid particle within the *k*-th calculation grid, the drag force was

$$\mathbf{F}_{p,\text{drag}}(t) = -\nabla P_{g,k} V_p + \frac{\beta_m V_p}{\varepsilon_{sm}} \left[\mathbf{u}_g(X_p) - \mathbf{u}_p \right] \tag{10}$$

where $\nabla P_{g,k}$ is the solid-phase pressure gradient within the *k*-th calculation grid, $\mathbf{u}_g(X_p)$ is the average velocity of the gas phase at the coordinate X_p, $V_p = W_p \frac{\pi D_p^3}{6}$ is the volume of the parcels, and β_m is the gas–solid phase energy transfer coefficient.

The gas–solid phase energy exchange of the *m*-th component in the *k*-th cell:

$$I_{gm}^k = \frac{1}{V_k} \sum \left[W_p^{(i)} F_{p,\text{drag}}^{(i)}(t) \right] \tag{11}$$

where V_k is the volume of the *k*-th cell.

The frictional stress between the particles is [20,21]

$$\mathbf{F}_{p,\text{coll}}(t) = -\frac{V_p}{\varepsilon_s} \nabla \tau \tag{12}$$

where $\nabla \tau$ is the interparticle pressure. The continuous particle corresponding force model based on Harris and Crighton [22] was selected to calculate the $\nabla \tau$:

$$\tau = \frac{P_s \varepsilon_s^\beta}{\max \left[\varepsilon_{s,cp} - \varepsilon_s^\beta, \varepsilon(1 - \varepsilon_s) \right]} \tag{13}$$

where P_s is a pressure constant (usually 1–100 Pa), $\varepsilon_{s,cp}$ is the close packing state volume fraction, β is a constant ($2 \leq \beta \leq 5$ is recommended), and ε is generally 10^{-7}, which was used to eliminate the singularity of the numerical calculation [23].

2.2. Chemical Reactions

Gas-solid fluidized combustion is a process containing a series of complex chemical reactions. The main components of the fuel in the fluidized bed are fixed carbon, volatiles, moisture, and inert ash. After the fuel enters the furnace, the water evaporates rapidly, the fuel particles are heated and volatilized, and the remaining coke particles react with oxygen or water vapor to burn. Ash stays longer in the furnace than other components, so the calculation of ash can be handled separately. The combustible gases of different compositions generated during the coke combustion and volatilization analysis processes are burned in oxygen. The combustion process of powder coke is shown in Figure 1.

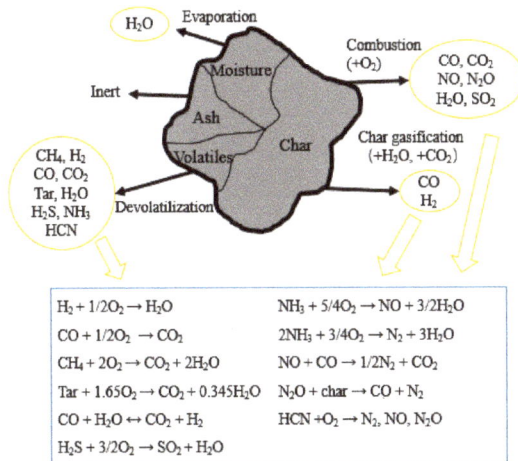

Figure 1 diagram labels and reactions:

- H$_2$O — Evaporation
- Combustion (+O$_2$) → CO, CO$_2$, NO, N$_2$O, H$_2$O, SO$_2$
- Moisture
- Inert ← Ash
- Char
- Char gasification (+H$_2$O, +CO$_2$) → CO, H$_2$
- CH$_4$, H$_2$, CO, CO$_2$, Tar, H$_2$O, H$_2$S, NH$_3$, HCN ← Volatiles — Devolatilization

H$_2$ + 1/2O$_2$ → H$_2$O	NH$_3$ + 5/4O$_2$ → NO + 3/2H$_2$O
CO + 1/2O$_2$ → CO$_2$	2NH$_3$ + 3/4O$_2$ → N$_2$ + 3H$_2$O
CH$_4$ + 2O$_2$ → CO$_2$ + 2H$_2$O	NO + CO → 1/2N$_2$ + CO$_2$
Tar + 1.65O$_2$ → CO$_2$ + 0.345H$_2$O	N$_2$O + char → CO + N$_2$
CO + H$_2$O ↔ CO$_2$ + H$_2$	HCN + O$_2$ → N$_2$, NO, N$_2$O
H$_2$S + 3/2O$_2$ → SO$_2$ + H$_2$O	

Figure 1. Main reactions of the powder coke combustion process.

In order to calculate the gas–solid phase heterogeneous reaction (such as coke combustion, limestone decomposition, desulfurization, denitration reaction, etc.), the average cell chemistry method was selected. In this method, the parameters of the discrete solid-phase particles are converted into intragrid parameters before the chemical reaction is calculated, and the consumption and generation of each component are calculated separately in each calculation grid.

There are thousands of reactions in a circulating fluidized bed boiler. It is almost impossible to calculate every reaction that occurs in the furnace. In order to simulate those reactions, such as devolatilization, volatile and char combustion, desulfurization, NO$_x$ formation, and denitration, only the major reactions have been taken into consideration. The corresponding reaction rates were given in our previous research [24].

3. Numerical Solution

In the MP-PIC method, the discretization of the solid phase is a major advantage over the two-fluid method. This statistical average and rediscrete processing balance the time consumption of the chemical reaction calculation while maintaining the accuracy of solid-phase flow simulation. In the actual operation process, when selecting the appropriate N_m, one can not only maintain the same calculation accuracy as the discrete element method (DEM) but also adapt to large-scale industrial simulation applications. The calculation process is shown in Figure 2 below.

After the calculation starts, the data is first initialized. Initialization includes not only initialization of the gas-phase velocity field, pressure field, density field, and temperature field but also the calculation of particle discretization. The velocity and pressure distribution of the gas phase are calculated by the SIMPLE algorithm [25]. For the calculation of the solid phase, the force and motion are analyzed for every parcel separately. After the gas–solid flow calculation is completed, the information on the discrete particle phase is then averaged inside the grid. A series of heterogeneous chemical reactions, such as particle combustion and gasification, are calculated based on the averaging treatment, and the homogeneous reactions of the gas phase are also calculated. After the chemical reaction calculation is completed, the updated grid information is rediscretized into a discrete distribution of calculated particles, and then the next gas–solid flow calculation is completed.

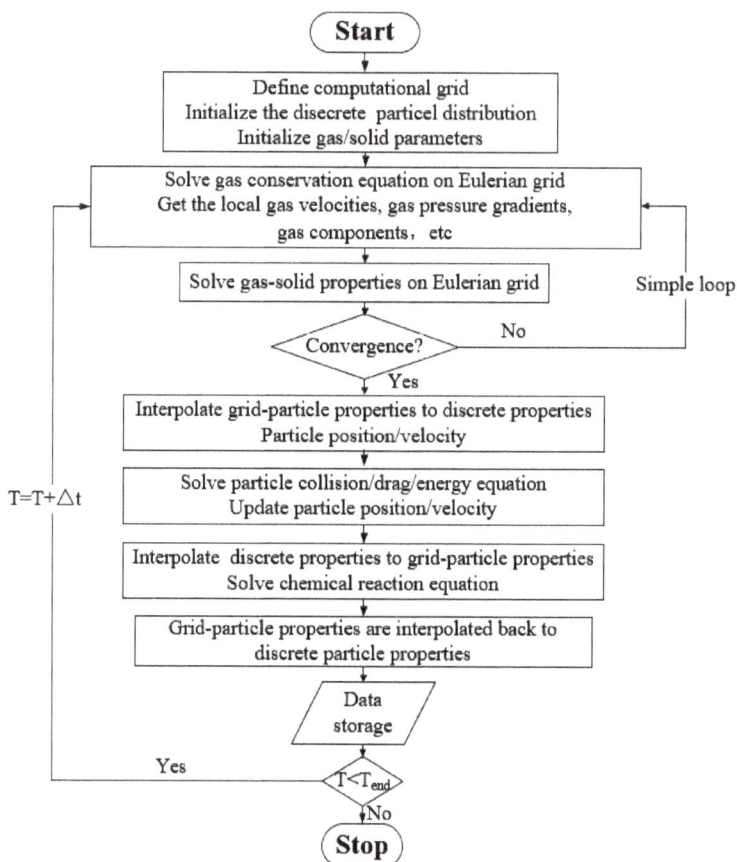

Figure 2. Calculation flow chart.

4. Simulation Object and Model Setup

Figure 3 shows the Dongfang boiler's experimental equipment. Figure 4 illustrates a schematic diagram of the experimental system, which contains a 24.5-m tall furnace, a cyclone, a recirculation bed, powder coke, and limestone feeding system. The main part of the fluidized bed system and the 3D grid of the furnace are shown in Figure 5. Detailed descriptions of the experiments can be found in Table 1. Table 2 is the elemental analysis of the powder coke used in the experiment. The particle size distribution of the powder coke is shown in Figure 6.

To investigate the impact of the operating parameters on the pollutant discharge characteristics of powder coke, this experiment used the L_9 (3^4) type orthogonal test method. The main design and operating parameters closely related to the fluidized bed combustion process (bed temperature, oxygen, and grading combustion) were investigated to provide important scientific guidance for managing and optimizing a powder coke CFB boiler.

Figure 3. Dongfang boiler 3-MW circulating fluidized bed combustion test bench.

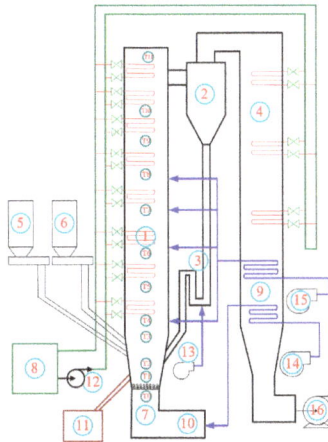

Figure 4. Schematic diagram of main equipment of the circulating fluidized bed (CFB) test bench. (1) Combustion chamber; (2) Cyclone separator; (3) Loop-seal recycle device; (4) Tail flue; (5) Fuel feeder; (6) Limestone feeder; (7) Fluidization air chamber; (8) Cooling water system (9) Air preheater; (10) Ignition unit; (11) Ash discharge system; (12) Feed pump; (13) High-pressure fluidized blower; (14) Primary fan; (15) Secondary air fan; (16) Draught fan.

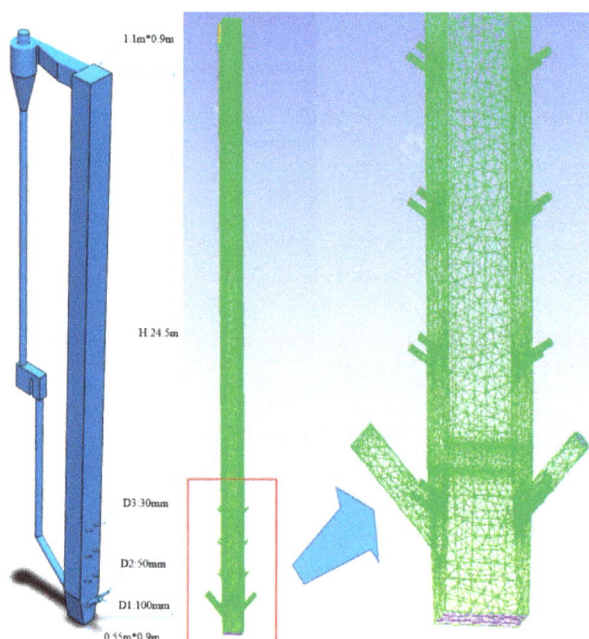

Figure 5. (**Left**) schematic of the CFB rig; (**right**) numerical grid of the furnace.

Table 1. Descriptions of the experiments.

	Units	Value
Thermal power (LHV)	MWth	3.0
Apparent velocity	m/s	~5
Cross-section area of dense phase	mm^2	550×900
Cross-section area of dilute phase	mm^2	1100×900
Height of the furnace	mm	24,500

Table 2. Analysis of powder coke.

Proximate Analysis	
Moisture (%)	19.92
Volatile matter (%)	8.83
Ash (%)	10.74
Fixed carbon (%)	60.51
Ultimate Analysis (Dry Ash Free)	
Carbon (%)	90.68
Hydrogen (%)	1.92
Oxygen (%)	5.83
Nitrogen (%)	1.11
Sulfur (%)	0.46
Lower calorific value (MJ/kg)	21.89

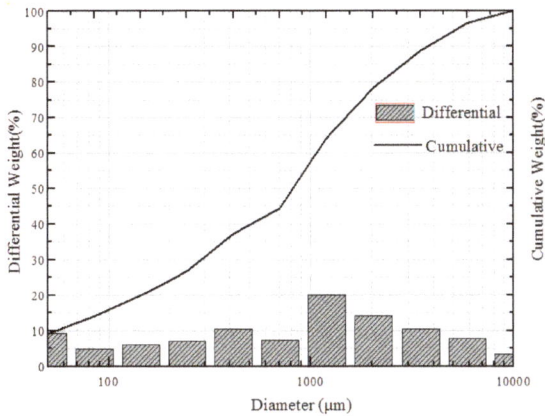

Figure 6. Particle size distribution of powder coke.

5. Results and Discussion

5.1. Determination of Fuel Ignition Point by TG-DTG Method

After the fuel particles were ground and subjected to thermogravimetric analysis on the US Perkin Elemer STA6000 instrument, the temperature–weightlessness–weight loss rate curve of the sample was obtained (Figure 7). The ignition point was determined by the thermal gravity-differential thermal gravity (TG-DTG) joint definition method. It can be seen from Figure 7 that with the increase in temperature, the TG curve below 120 °C had a rapidly declining trend, which was the rapid evaporation process of moisture. When the temperature exceeded 260 °C, the TG curve showed a second downward trend, which corresponded to the precipitation of volatiles, while some low-flammable volatiles began to ignite. The turning point of the TG curve was the ignition temperature, which was about 470 °C. As the temperature rose to 780 °C, the TG curve became a straight line, indicating that the fuel had burned out.

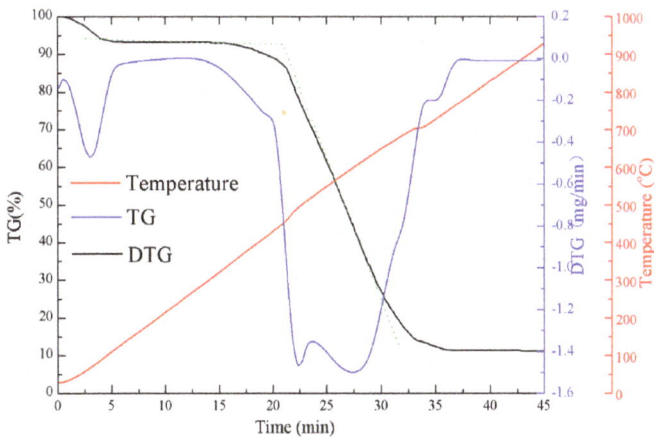

Figure 7. Diagram of ignition point determined by the powder coke thermal gravity-differential thermal gravity (TG-DTG) combined method.

5.2. Simulation Results of the First Set of Experimental Conditions

For the sake of verifying the accuracy of the 3D computational fluid dynamics (CFD) model based on the MP-PIC method, a set of stable operating conditions were selected for simulation. The main operating parameters of this working conditions are shown in Table 3. The calculated results were compared with the experimental results to verify the accuracy of the model.

Table 3. Experimental and modeling conditions.

	Units	Value
Fluidized air flow	m^3/s	0.42
Fluidized air pressure	Pa	110,955
Fluidized air temperature	°C	243.4
Second air flow	m^3/s	0.416
Second air pressure	Pa	106,325
Second air temperature	°C	198.8
Feed powder coke	Kg/s	0.03
Limestone	Kg/s	0
Average bed temperature	°C	916

Figures 8–10 show the main gas–solid parameters obtained by simulation. It can be seen from Figure 8 that there was a general tendency to move downward due to the injection of fuel particles and the backmixing of the large particle size from the return port at the bottom of the furnace. As the height increased, the fuel particles underwent a drying and crushing action, and the particle size gradually decreased. The movement of the particles in the 2.5-m dense phase region was complicated. When the height reached about 5 m, particles in the core area had an overall upward trend, but the particles near the wall moved downward along the water wall, showing a typical "core-annulus" shape. The particles in the core area flowed upward under the action of drag. Some of the particles continuously migrated to the vicinity of the wall. When the height exceeded 10 m, reaching the dilute phase region of the furnace, the backmixing between the core rings was gradually weakened, and the small particles in the dilute phase region tended to move upward. Figure 9 shows the distribution of the solid-phase volume fraction along the height. It can also be seen that the region below about 5 m was a dense phase region, and the solid-phase fraction was gradually reduced at 5 m or more.

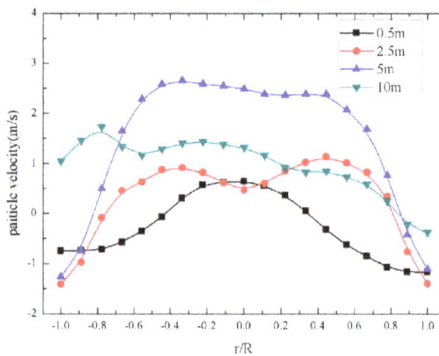

Figure 8. Radial distribution of particle velocities at different heights.

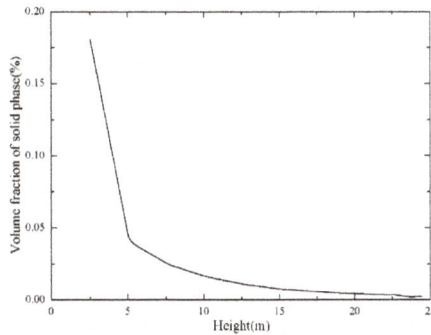

Figure 9. Distribution of solid volume fraction along the height.

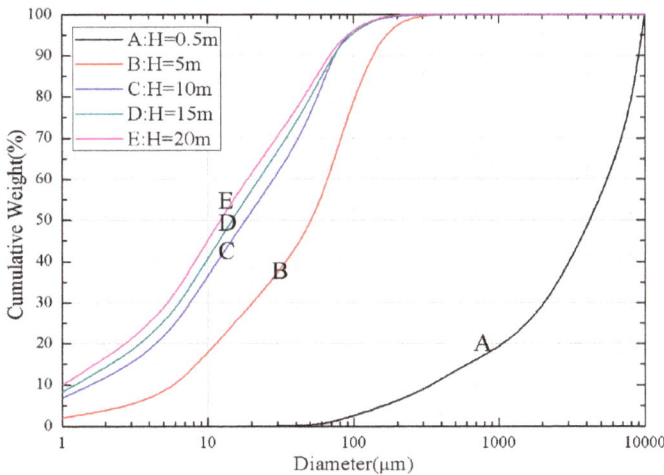

Figure 10. Particle size distributions at different heights.

The curves A–E in Figure 10 are distributions of the particle size distributions of the solid particles at heights of 0.3, 5, 15 and 20 m, respectively, in the furnace. As the height of the furnace increased, the proportion of small-sized particles became larger and larger due to the influence of solid particle precipitation and backmixing, while the proportion of large-sized particles became smaller and smaller. After some of the particles entered the cyclone with the flue gas, the particles that had undergone cyclone separation were returned to the furnace through the return sealing device. Large-sized particles were mostly concentrated in the dense phase region at the bottom of the furnace.

Figure 11 shows the comparison of temperature distribution under stable operating conditions between experiments and simulation. The temperature distributions obtained by simulation and experiment were quite close. After the coke was rapidly separated into volatiles after entering the furnace, the remaining coke was burned in the lower layer of the furnace to release a large amount of heat, resulting in the temperature of the lower layer of the furnace being relatively high. With the increase in furnace height, the oxygen concentration decreased with the consumption in the area below the secondary tuyere, and the temperature of the upper dense phase zone decreased with the addition of low-temperature fuel and limestone particles. At about 5 m, the fresh air brought from the secondary tuyere caused an increase of oxygen concentration, and the volatile combustion increased, forming a locally higher temperature. The temperature gradually decreased due to the cooling of the furnace wall above the secondary air vent.

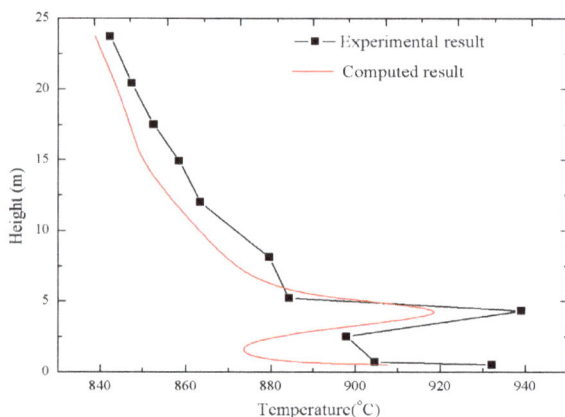

Figure 11. Comparison of temperature distribution under stable operating conditions.

As can be seen from Figure 12, the concentration of oxygen contained in the primary air at the inlet of the furnace gradually decreased after the gas–solid phase heterogeneous reaction occurred with the fuel particles in the dense phase region. The oxygen brought about by the secondary air caused a transient rise in the oxygen concentration near the secondary tuyere, and then the oxygen concentration gradually decreased as the gas-phase reaction proceeded. The changing trend of CO_2 concentration was just opposite to that of O_2, and it had been increasing after a short decline near the secondary air outlet. From the trend of CO concentration, it can be seen that in the dense phase region at the bottom of the furnace, the rate of CO formation was much larger than the consumption rate. Due to the insufficient oxygen concentration and the high concentration of combustibles in the dense phase region, the combustion was inadequate and a large amount of CO was generated. In the dilute upper phase of the furnace, the concentration of CO was gradually reduced, mainly because the consumption of combustibles was exhausted, and the remaining O_2 in the flue gas reacted with CO to form CO_2.

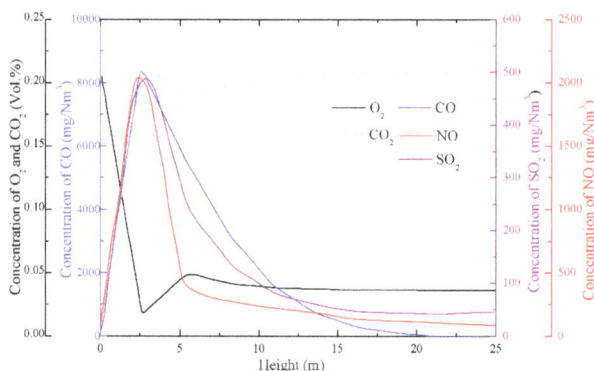

Figure 12. Distribution of gas concentrations along the height from the simulation.

Generally, sulfur in coal can be divided into two parts: organic sulfur and pyrite sulfur, in addition to a small amount of sulfate sulfur. Since the fluidized bed combustion temperature was generally controlled below 1000 °C, the sulfur present in the fuel as sulfate was not SO_2, so the amount of SO_2 produced by fluidized bed combustion was lower than that of a pulverized coal furnace. At the same time, the coal minerals in the fluidized bed boiler could absorb the SO_2 released by the combustion process. This characteristic of the fluidized bed combustion process is called the inherent desulfurization capacity of coal (i.e., the self-desulfurization capacity). Due to the low-temperature combustion characteristics of the fluidized bed combustion process, the addition of a desulfurizing agent (e.g., an alkaline earth metal oxide such as CaO) in the furnace absorbed SO_2 to form a stable sulfate compound, thereby realizing the possibility of efficient desulfurization. The main factors affecting the desulfurization efficiency of a CFB boiler include the Ca/S molar ratio, operating bed temperature, fuel particle size, operating oxygen, staged combustion (primary air rate and secondary air ratio), SO_2 residence time in the furnace, circulation material quantity, and load change [26]. It can be seen from Figure 12 that the concentration of SO_2 gradually increased with the increase of the height in the dense phase region, mainly because the sulfur element in the fuel in the bottom region of the furnace was gradually released, and a large amount of SO_2 was formed. Since powdered coke fuel has a molar ratio of calcium to sulfur of 5.66, the powder coke had strong self-desulfurization performance. As the height increased, a large amount of SO_2 was absorbed by the fuel's own CaO to form $CaSO_4$. When no additional desulfurizer was added, the SO_2 concentration at the exit of the furnace was only 45 mg/Nm^3.

There are two main NO_x generation pathways. One is that the nitrogen element in the air reacts with oxygen at a high temperature to form NO_x, that is, thermal NO_x. The other is that the nitrogen element in the fuel is thermally decomposed and oxidized to form NO_x, that is, fuel-type NO_x. The NO formed by fuel nitrogen accounts for more than 95% of the total NO_x emission of the fluidized bed combustion mode [27]. It was assumed that the main component of NO_x is NO, and the effect of N_2O on NO_x emissions was ignored in these simulations. When the coal was burned, part of the combustion was analyzed by volatilization (i.e., volatile nitrogen), and the other part remained on coke (i.e., coke nitrogen). Studies have shown that the main factors affecting the NO_x emissions of circulating fluidized bed boilers are operating temperature, excess air coefficient, staged combustion, desulfurizer, fuel properties, cycle rate, selective noncatalytic reduction (SNCR), and ammonia injection [26]. It also can be seen from Figure 12 that after entering the furnace, the fuel particles rapidly generated a large amount of fuel-type NO_x in a high-temperature environment, and then the NO_x concentration gradually decreased with the reduction of CO and the dilution of the secondary air. The NO_x concentration at the exit of the furnace was only about 90 mg/Nm^3.

The highest concentration of pollutants occurred near the fuel feed point. When there was no limestone desulfurization and SCR/SNCR denitrification, the total discharge of pollutants from the furnace outlet was only about 135 mg/Nm^3, which shows that the CFB boiler has a relatively low rate of pollutant emissions.

5.3. Factors Affecting Pollutant Emissions

Based on the above experiments and simulation results, it can be found that the flow combustion parameters' distribution in the test bench was very similar to that of the actual power plant fluidized bed. Therefore, a series of studies on the factors affecting the actual power plant emissions were carried out on this test bench. Table 4 shows the powder coke burning test conditions.

Table 4. Powder coke burning test condition.

	Average Bed Temperature (°C)	Operating Oxygen (%)	Primary Air Ratio (%)
1	916	3.0	51
2	916	3.7	46
3	916	4.2	40
4	890	3.0	46
5	890	3.7	40
6	890	4.2	51
7	965	3.0	40
8	965	3.7	51
9	965	4.2	46

Figure 13a–c shows three factors affecting the discharge of pollutants, which are operating bed temperature, operating oxygen, and primary air rate. Because the temperature, pressure, and moisture in the flue gas of different fuels and boilers are different, for the convenience of horizontal comparison, the concentrations of pollutants were converted to 6% O_2, and dry flue gas concentration under standard conditions was used (1 atm and 0 °C).

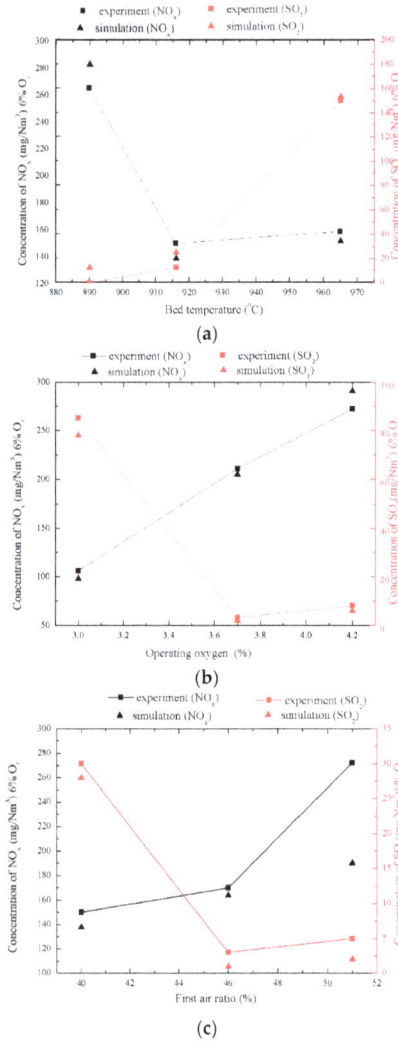

(a)

(b)

(c)

Figure 13. Three factors affecting pollutant emissions. (**a**) Effect of bed temperature on pollutant emissions; (**b**) Effect of operating oxygen on pollutant emissions; (**c**) Effect of first air ratio on pollutant emissions.

It can be seen in Figure 13a that the increase of bed temperature will significantly affect the emission of SO_2. As the bed temperature increases, the self-desulfurization ability of the fuel decreases, resulting in a sharp rise in SO_2 emission concentration. The relationships between SO_2 emission, operating oxygen, and primary air rate are shown in Figure 13b,c. The change trends of these two curves are quite similar. When the operating oxygen is 3.0% and the primary air rate is 40%, the SO_2 emission is the highest. With the increase of operating oxygen and primary air rate, SO_2 emissions are gradually reduced. After the operating oxygen exceeds 3.6% and the primary air rate exceeds 46%, the changes of the two have little effect on SO_2 emissions.

It also can be seen from Figure 13 that the factors affecting the NO_x emission concentration during the normal bed temperature and stable combustion of the powdered coke fuel are from large to small in the order of operating oxygen > operating bed temperature > primary air rate. After the bed temperature exceeds 916 °C, the generation of NO_x does not change substantially. Similarly, the amount of NO_x generated during the increase of primary air rate from 40% to 46% is also small, but the change of NO_x emission and operating oxygen quantity is linear. The NO_x emission concentration value of the powdered coke fuel under normal bed temperature and stable combustion conditions increases with the increase of oxygen content and primary air rate. Considering the influence of bed temperature on SO_2 emissions, high temperatures (such as 965 °C) and low temperatures (such as 890 °C) are not conducive to reduce pollutant discharge in fluidized bed powder coke combustion; the optimal operating bed temperature is about 916 °C. After the L_9 (3^4) type orthogonal table test data processing, it was found that the optimal test conditions for making the NO_x emission concentration the lowest are a bed temperature of 916 °C, an oxygen content of 3.0%, and a primary air rate of 40%.

The comparison between the simulation results of these nine test conditions and the experimental results shows that the grid-based average chemical reaction calculation and the MP-PIC-based gas–solid flow simulation have relatively high precision for powder coke fluidized bed combustion and pollutant discharge. However, when the primary air rate was 51%, the NO_x emission in the simulation results showed a certain error compared with the experimental results, but the overall change trend was basically correct, which indicates that the calculation model still has room for improvement.

6. Conclusions

As a new type of fluidized bed fuel, powder coke, which is carbonized by coal, has been studied in this paper for its fluidized combustion and pollutant emission characteristics. The factors affecting the stability and pollutant emission of powder coke in the furnace combustion of a fluidized bed boiler have been studied through a series of experiments. At the same time, a 3D CFD model of the furnace based on the MP-PIC method has been developed. The particle size distribution of the particles was taken into consideration in this model. The particle packing form was used to describe the solid phase to reduce the quantity of calculations. The discrete gas–solid heterogeneous reactions were averaged in the grid cell. The simulated results were compared with the experimental results. The calculated temperature distribution in the furnace was similar to the experimental results, the maximum error was less than 25 °C, and the porosity distribution and solid-phase velocity distribution in the furnace were correct. The distribution of gases was similar to that of other fluidized bed boilers [24]. The results show that the model has fairly high precision and can be used for further research.

The NO_x pollutant emission value of the normal bed temperature stable combustion test conditions ranged from 70.5 to 507.9 mg/Nm^3 (dry flue gas, 6% oxygen). The analysis of NO_x emission characteristics of multiple test conditions showed that CFB was used. The NO_x emission value was affected by various factors, such as operating bed temperature, oxygen volume, and primary air rate, during combustion. The self-desulfurization characteristics in the CFB furnace were mainly affected by fuel type, self-calcium ratio of calcium to sulfur, and boiler operation mode. For powder coke fuel fluidized combustion, operating oxygen ($\leq$3.0%) and primary air rate ($\leq$40%) were low and the average operating bed temperature was high ($\geq$960 °C), which seriously affected the self-desulfurization capacity of the powder coke fuel and led to an increase in SO_2 emission concentration in the operation of oxygen from high to low (such as from 4.2% to 3.0% or less.) Since the SO_2 emission concentration is highly likely to rise sharply, it is therefore recommended to avoid the above operating conditions during CFB boiler operation and to design a process to improve the powder coke fuel furnace. The self-desulfurization ability in the internal combustion process enables the implementation of desulfurization agents such as limestone and an SO_2 emission value that meets the current strict SO_2 emission standards.

This article mainly introduced the combustion and pollutant emission test of powder coke in the fluidized bed boiler test bench. At the same time, we developed a 3D CFD model based on the MP-PIC method for a fluidized bed boiler fluidization reaction system to investigate the internal gas–solid flow and reactions. As a special fluidized bed boiler fuel, powder coke has a different linearity of composition and pollutant emissions than traditional coal. In future research, we will try to extend this method to the application of a large-scale commercial 600-MW circulating fluidized bed. In the near future, we will combine the research results of the core-annulus cell model to simulate the Baima Demonstration Power Station, further verify the accuracy and stability of the MP-PIC method, and provide technical guidance for future plant design, operation, and operations based on the calculation results.

Author Contributions: Conceptualization, H.W.; Formal analysis, L.S.; Funding acquisition, C.Y.; Investigation, H.W.; Methodology, H.W.; Project administration, C.Y.; Resources, H.H.; Supervision, C.Y.; Visualization, K.D.; Writing—original draft, H.W.; Writing—review & editing, H.W.

Funding: This research was funded by the National Natural Science Foundation of China, grant numbers 51576020 and 51876011.

Conflicts of Interest: The authors declare no conflicts of interests.

Nomenclature

ε	Void fraction
h	Height of furnace (m)
ρ	Density (kg/m^3)
g	Gravity (m/s^2)
S	Mass changes of each component due to the chemical reaction (kg/m^3/s)
$\mathbf{u}$	Velocity of gas and solid phase (m/s)
d	Diameter of particle (m)
h	Convective heat transfer coefficient (W/(m^2·K))
t	Time (s)
p	Pressure (Pa)
C_p	Specific heat capacity (J/kg·K)
T	Temperature (K)
μ	Viscosity (Pa·s)
λ	Thermal conductivity (W/(m·K))
C	Mole concentration of each species
Re	Reynolds number
Sh	Sherwood number
Φ	Viscous dissipation
$\dot{Q}$	Energy source
$\dot{q}_D$	Enthalpy diffusion term
m	Mass (kg)
$\mathbf{F}_{p,\text{drag}}(t)$	Drag force of the parcel
$\mathbf{F}_{p,\text{coll}}(t)$	Interaction force between the parcels
$\nabla P_{g,k}$	Solid-phase pressure gradient
β_m	Gas–solid phase energy transfer coefficient
V	Volume
X	Position
Subscripts	
i	Number of gas-phase species
h	Height
g	Gas phase
s	Solid phase
k	k-th cell
cp	Close packing state
p	Properties of parcel

References

1. Zhuang, Y.-Q.; Chen, X.-M.; Luo, Z.-H.; Xiao, J. CFD–DEM modeling of gas–solid flow and catalytic MTO reaction in a fluidized bed reactor. *Comput. Chem. Eng.* **2014**, *60*, 1–16. [CrossRef]
2. Gerber, S.; Behrendt, F.; Oevermann, M. An Eulerian modeling approach of wood gasification in a bubbling fluidized bed reactor using char as bed material. *Fuel* **2010**, *89*, 2903–2917. [CrossRef]
3. Leckner, B.; Gómez-Barea, A. Oxy-fuel combustion in circulating fluidized bed boilers. *Appl. Energ.* **2014**, *125*, 308–318. [CrossRef]
4. Agung, A.; Lathouwers, D.; van der Hagen, T.H.; van Dam, H.; Pain, C.C.; Goddard, A.J.; Eaton, M.D.; Gomes, J.L.; Miles, B.; de Oliveira, C.R. On an improved design of a fluidized bed nuclear reactor-I: Design modifications and steady-state features. *Nucl. Technol.* **2006**, *153*, 117–131. [CrossRef]
5. Gidaspow, D. Hydrodynamics of fiuidizatlon and heat transfer: Supercomputer modeling. *Appl. Mech. Rev.* **1986**, *39*, 1–23. [CrossRef]
6. Chen, S.; Fan, Y.; Yan, Z.; Wang, W.; Lu, C. CFD simulation of gas–solid two-phase flow and mixing in a FCC riser with feedstock injection. *Powder Technol.* **2016**, *287*, 29–42. [CrossRef]
7. Liu, D.; Chen, X.; Zhou, W.; Zhao, C. Simulation of char and propane combustion in a fluidized bed by extending DEM–CFD approach. *Proc. Combust. Inst.* **2011**, *33*, 2701–2708. [CrossRef]
8. Sinclair, J.; Jackson, R. Gas-particle flow in a vertical pipe with particle-particle interactions. *AIChE J.* **1989**, *35*, 1473–1486. [CrossRef]
9. Pita, J.A.; Sundaresan, S. Gas-solid flow in vertical tubes. *AIChE J.* **1991**, *37*, 1009–1018. [CrossRef]
10. Snider, D.M.; Clark, S.M.; O'Rourke, P.J. Eulerian–Lagrangian method for three-dimensional thermal reacting flow with application to coal gasifiers. *Chem. Eng. Sci.* **2011**, *66*, 1285–1295. [CrossRef]
11. Xie, J.; Zhong, W.; Jin, B.; Shao, Y.; Huang, Y. Eulerian–Lagrangian method for three-dimensional simulation of fluidized bed coal gasification. *Adv. Powder Technol.* **2013**, *24*, 382–392. [CrossRef]
12. Snider, D.M.; O'Rourke, P.J.; Andrews, M.J. Sediment flow in inclined vessels calculated using a multiphase particle-in-cell model for dense particle flows. *Int. J. Multiph. Flow* **1998**, *24*, 1359–1382. [CrossRef]
13. Snider, D.; Banerjee, S. Heterogeneous gas chemistry in the CPFD Eulerian–Lagrangian numerical scheme (ozone decomposition). *Powder Technol.* **2010**, *199*, 100–106. [CrossRef]
14. Abbasi, A.; Ege, P.E.; De Lasa, H.I. CPFD simulation of a fast fluidized bed steam coal gasifier feeding section. *Chem. Eng. J.* **2011**, *174*, 341–350. [CrossRef]
15. Ryan, E.M.; DeCroix, D.; Breault, R.; Xu, W.; Huckaby, E.D.; Saha, K.; Dartevelle, S.; Sun, X. Multi-phase CFD modeling of solid sorbent carbon capture system. *Powder Technol.* **2013**, *242*, 117–134. [CrossRef]
16. Adamczyk, W.P.; Węcel, G.; Klajny, M.; Kozołub, P.; Klimanek, A.; Białecki, R.A. Modeling of particle transport and combustion phenomena in a large-scale circulating fluidized bed boiler using a hybrid Euler–Lagrange approach. *Particuology* **2014**, *16*, 29–40. [CrossRef]
17. Loha, C.; Chattopadhyay, H.; Chatterjee, P.K. Three dimensional kinetic modeling of fluidized bed biomass gasification. *Chem. Eng. Sci.* **2014**, *109*, 53–64. [CrossRef]
18. Zhong, W.; Xie, J.; Shao, Y.; Liu, X.; Jin, B. Three-dimensional modeling of olive cake combustion in CFB. *Appl. Therm. Eng.* **2015**, *88*, 322–333. [CrossRef]
19. Gidaspow, D. *Multiphase Flow and Fluidization: Continuum and Kinetic Theory Descriptions*; Academic Press: Cambridge, MA, USA, 1994.
20. O'Rourke, P.J.; Snider, D.M. An improved collision damping time for MP-PIC calculations of dense particle flows with applications to polydisperse sedimenting beds and colliding particle jets. *Chem. Eng. Sci.* **2010**, *65*, 6014–6028. [CrossRef]
21. O'Rourke, P.J.; Zhao, P.P.; Snider, D. A model for collisional exchange in gas/liquid/solid fluidized beds. *Chem. Eng. Sci.* **2009**, *64*, 1784–1797. [CrossRef]
22. Harris, S.; Crighton, D. Solitons, solitary waves, and voidage disturbances in gas-fluidized beds. *J. Fluid Mech.* **1994**, *266*, 243–276. [CrossRef]
23. Auzerais, F.; Jackson, R.; Russel, W. The resolution of shocks and the effects of compressible sediments in transient settling. *J. Fluid Mech.* **1988**, *195*, 437–462. [CrossRef]
24. Wu, H.; Yang, C.; He, H.; Huang, S.; Chen, H. A hybrid simulation of a 600 MW supercritical circulating fluidized bed boiler system. *Appl. Eng.* **2018**, *143*, 977–987. [CrossRef]
25. Patankar, S. *Numerical Heat Transfer and Fluid Flow*; CRC Press: Boca Raton, FL, USA, 1980.

26. Scala, F. *Fluidized Bed Technologies for Near-Zero Emission Combustion and Gasification*; Elsevier: Amsterdam, The Netherlands, 2013.

27. Wang, Q.; Luo, Z.; Li, X.; Fang, M.; Ni, M.; Cen, K. Modeling of NO and N2O formation and decomposition in circulating fluidized bed boiler. *J. Fuel Chem. Technol.* **1998**, *26*, 108–113.

energies

MDPI

Article

Efficiency of the Air-Pollution Control System of a Lead-Acid-Battery Recycling Industry

Kyriaki Kelektsoglou *, Dimitra Karali, Alexandros Stavridis and Glykeria Loupa

Department of Environmental Engineering, Democritus University of Thrace, 12 Vas. Sofias,
Xanthi 67100, Greece; dkarali@env.duth.gr (D.K.); stavridisa@gmail.com (A.S.); gloupa@env.duth.gr (G.L.)
* Correspondence: kkelekt@env.duth.gr

Received: 31 October 2018; Accepted: 6 December 2018; Published: 11 December 2018

Abstract: The air-pollution control system of a lead-acid-battery recycling industry was studied. The system comprised two streams with gravity settlers followed by filter bags for the factory indoor air and the metal-recycling furnace, respectively. Efficiency in particle removal according to mass was found to be 99.91%. Moreover, filter bags and dust from the gravity settlers were analyzed for heavy metals by Wavelength Dispersive X-Ray Fluorescence. The results showed high concentrations of Pb and Na in all cases. In the filter bag samples from the indoor atmosphere stream, Ca, Cu, Fe, and Al were found in concentrations higher than that in the filter bag samples from the furnace stream. The opposite was found for Na. Tl and K were only found in furnace stream bag filters. The elemental concentration of the dust from the furnace fumes stream contained mainly Fe, Na, Cd, Pb, Sb, and Cl, while the indoor main stream contained mainly P, Fe, Na, Pb, and Sb. In all cases, impurities of Nd, Ni, Rb, Sr, Th, Hg, and Bi were found. The high efficiency of the air-pollution control system in particle removal shows that a considerable reduction in emissions was achieved.

Keywords: air-pollution control; battery recycling; heavy metals; control system efficiency; chemical analysis

---

1. Introduction

Battery recycling is the activity that aims to reduce the number of batteries being disposed as solid waste. Batteries contain heavy metals and toxicants that can contaminate soil and water. Spent batteries from cars, motorcycles, uninterruptible power source UPS, marine applications, fork-lift trucks, electric vehicles, and many others represent an important secondary source of metals (mainly lead) in very high concentration levels, sometimes higher than in natural sources; these metals sometimes are very expensive. Lead recycling saves energy since it is far more energy-efficient to recycle than it is to produce lead from mining and processing ores. Therefore, the battery-recycling approach is a sustainable process for both natural and economic resources.

Outdated recycling processes in some developing countries result in large amounts of lead-dust fumes, dust, and hazardous wastes that cause serious problems to human's health [1–3]. Dust particles emitted indoors by furnaces and metallurgical smelters are enriched in metallic compounds, and this may be deleterious to workers. Heavy metals cause serious problems to human' health [4] and should be kept out of the waste stream. Consequently, the operation of an appropriate pollution control system in such industries is essential. There are only a few studies in the field of battery recycling that are occupied with its air pollution control system [5,6]. Rada et al. [5] and Pan et al. [6] proposed a new recycling method that fulfilled the requirements of an eco-innovative technology controlling air pollution. Due to concerns that pollutants are captured by the individual parts of an air-pollution control system, there have been several studies [7–20]. Only the studies conducted by Sobanska et al. [19] and Spear et al. [20] investigated the fly ash and dust emitted by primary lead smelters. Limited information is available on the quantitative distribution of heavy metals emitted by the battery recycling process and captured by an air-pollution control system. According to the literature, only Ettler et al. [21] investigated the mineralogy of air-pollution control system residues from a car-battery recycling center in the Czech Republic.

The objective of this study was to investigate the air-pollution control-system efficiency of a lead-battery recycling factory and to analyze the heavy-metal concentrations from different origins in the air-pollution control system, such as in the bag house filters and the dust from the tanks that are gathered by gravity settlers and bag houses.

A lead-acid battery recycling factory, "Sunlight", in Greece was investigated. Sunlight recycling is the lead-acid battery recycling branch of systems SUNLIGHT S.A., which has an area of 4.2 hectares and it is located in the Industrial Area of Komotini, Thrace, Northern Greece. Lead-acid batteries are mainly car and motorcycle batteries, industrial fork-lift batteries, and UPS batteries. The recycling cycle consists of the collection of the batteries, breaking batteries into smaller pieces, neutralization of the acid content, separation of the metal (Pb) grid from the plastic casing, cleaning of the metal grid, and melting the metal grid in a high temperature furnace for recycling or further metallurgical processes. The factory is able to recycle 25,000 tons of spent batteries per year, resulting in financial and environmental benefits.

2. Materials and Methods

Air-Pollution Control-System Description

In Figure 1, the Sunlight air-pollution control system is depicted. The furnace KL-710 and its produced fumes are under ventilation by Fan U-720, which regulates the flow rate up to 40,000.00 m textsuperscript3/h. Polluted air from the furnace is first brought through flue-dust gravity settling chamber MC 720 and then through bag-house PK-720. The latter consists of 792 filtering bags, Dn 123/3100 mm made of acrylic co-polymers with a waterproof membrane of 600 g/m^2, suitable for continuous operation up to 130 °C.

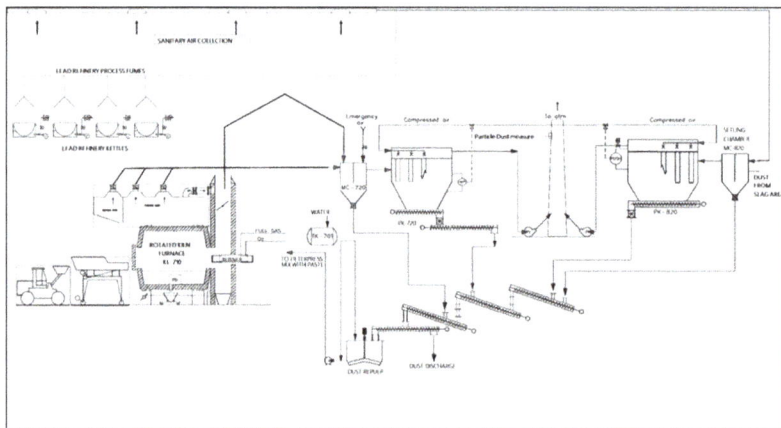

Figure 1. Sunlight air-pollution control system.

Moreover, the sanitary air from the lead refinery kettles and the surmounted space is ventilated by a U-820 fan with a flow rate up to 120,000.00 m^3/h. The collected stream is first brought through another flue-dust gravity settling chamber, MC 820, and then sent to a PK820 bag house that consists of 1188 filtering bags, Dn 123 × 4000 mm, of acrilyc needle felt with surface treatment anti-block waterproof PTFE 600 g/m^2, suitable for continuous operation up to 140 °C.

As dust collects on the filters, a dust layer builds up that increases the pressure inside the bag, hence increasing the energy requirements to move the gas through the dust cake. Eventually, the dust layer becomes so thick that the pressure threshold is exceeded and the dust cake needs to be removed. The type of dust-cake removal (cleaning) in our case is pulse-jet self-cleaning. The air supply used for cleaning is almost 600 kPa and is applied when the pressure inside the bag exceeds 2 kPa.

The flue dust selected at the bottom of the two bag houses is transported by the Archimedes screw conveyor systems into the flue-dust tank V-710. The dust from the bottom of the two gravity settling chambers is also discharged to the same tank. The dust is suspended in the tank with water and then it is filtered to a paste. The paste is sent back to the furnace in order to reduce the volume and the amount of fine dust in the working area.

The "free"-of-particles air stream after the bag houses (up to 160,000.00 m^3/h) is emitted through a self-lifting stack into the atmosphere. The stack dimensions are 20 m height and 2.5 m diameter.

3. Results/Discussion

3.1. Air-Pollution Control-System Efficiency in Particle Removal

The calculation of the system's efficiency was realized for a time period of about three months from April 2018, when the bag house filters were replaced by clean new filters, to June 2018, when samples from the bag filters were selected and analyzed. The bags are cleaned by intermittent jets of compressed air that flow into the inside of the bag to blow the cake off. Often, these bag houses are cleaned while they are in service; the internal pulse causes many of the collected solids to fall to the hopper, but some remain in the fabric of the filter cloth.

According to the manufacturer's specifications, the total flow rate in the stack was up to 160,000.00 m textsuperscript3/h. The flow rates reaching the stack after PK720 and PK820 were measured, and the real flow rate in the stack was calculated by their addition. There were continuous measurements for the whole time period that we conducted this study. There were also data from the particle concentration (mg/m^3) measured at the same time in the stack. The mass of the particles that were not captured by the control system m_{out} was obtained by multiplying the real flow rates with particle concentration at every moment. The total m_{out} was found by adding all these values for the period of three months and was 167.7 kg. Furthermore, the particles captured by the flue dust system at the same period was gathered in the tank and was weighed (m_{dust}). It was 190,862 kg. The m_{filter} is the mass captured by the filter and obtained by gravimetry. The bag filters that were used for heavy-metal analysis were weighed. The blanks were also pre-weighed. Subtracting the blank from the bag filter sample, the weight of the captured dust was found. There were six filters (each one 12.56 cm^2) from every bag house, and we calculated the mean dust captured from these filters. It was 0.14095 gr for PK820 and 0.1483 for PK720. As the total filtering surface was 1820 and 964 m^2 for PK820, and PK720, respectively, and also considering homogeneity to the whole filter, we calculated 209 kg of captured dust for PK820, and 114 kg for PK720. Adding these amounts, we found the total dust captured by the bag filters $\left(m_{filter} \right)$ to be 232 kg. The sum of m_{dust} and m_{filter} gives the total mass captured by the control system.

As a result, the particle mass reaching the air-pollution control system in Equation (1) was found to be:

$$m_{in} = m_{dust} + m_{filter} + m_{out} \tag{1}$$

According to Equation (2), the total efficiency of the system was:

$$n = \frac{m_{in} - m_{out}}{m_{in}} \tag{2}$$

For
$m_{dust} = 190{,}862$ kg

$m_{filter} = 323$ kg

$m_{out} = 167.7$ kg

The total n was calculated to be 99.91%.

According to Liu et al. [22], the weight efficiency for the total dust in membrane filters in bag houses is 99.9% and this is in agreement with our results.

Moreover, the efficiency–diameter relation for the two gravity settlers for both the blocked (n_b) (Laminar flow) and mixed (n_m) flow (Turbulent flow) models, assuming Stoke's law, was calculated. We assumed particle diameters ranging from 0.5 to 400 μm. In Figure 2a,b the efficiency–diameter relations for gravity settling of MC720 and MC820, respectively, are indicated. The flow rate from MC720 was assumed to be 40,000 m³/h and from MC820, 120,000 m³/h. In both cases, for small particles, for which the calculated collection efficiencies are small, the mixed and blocked flow models gave practically the same answers. For larger particles, the efficiencies became larger, and the two models gave different answers. For MC720, collection efficiency was more than 50% for particles larger than 200 μm in both flow models and reached the maximum of 80% for particles with a diameter larger than 400 μm mainly for a mixed flow. For MC820, the trend was almost the same. It reached the maximum collection efficiency of 60% for particles with a diameter larger than 400 μm for a mixed flow and almost 90% for block flow. Consequently, these two devices only showed higher efficiency for coarse particles larger than 200 μm and this results in the need for the bag houses after the gravity filters. Generally, gravity settlers are old unsophisticated devices that are used in industries treating very dirty gases in order to remove, as a first step, the large particles [23,24]. They are followed, in most cases, by bag filters.

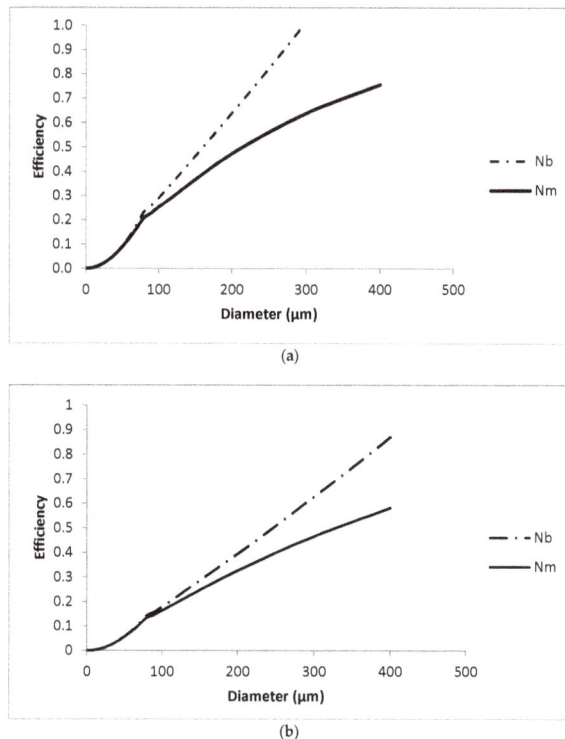

(a)

(b)

Figure 2. Efficiency-diameter relation for gravity settling in (**a**) MC720 (40,000 m³/h) and (**b**) MC820 (120,000 m³/h).

3.2. Particles in the Atmosphere

The suspended particles in the stack from April 2018 to 21 June 2018 were measured with a Laser Dust Monitor (NEOM), NEO Monitors AS, Norway. This is an optical instrument based on transmitting visible laser light from a transmitter unit on one side of the stack to a receiver unit on the diametrically opposite side of the stack. The measuring technique is based on measuring absorption and the scattering of light created by the dust particles present in the stack. The measurement signal corresponds to the integrated dust concentration over the entire optical path (stack). However, the instrument does not determine particles below "Aitken" nuclei. This is not significant for the total mass, but their presence in the atmosphere is important for human health. Particle concentration is indicated in Figure 3. According to the European Commission implementing EU Decision 2016/1032, establishing the best available techniques under directive 2010/75/EU, the upper limit of suspended particles is 5 mg/m^3 at the outflow of the chimney. In Figure 3, it is indicated that during the study period the measurements from the outflow concentration were below the upper limit of 5 mg/m^3.

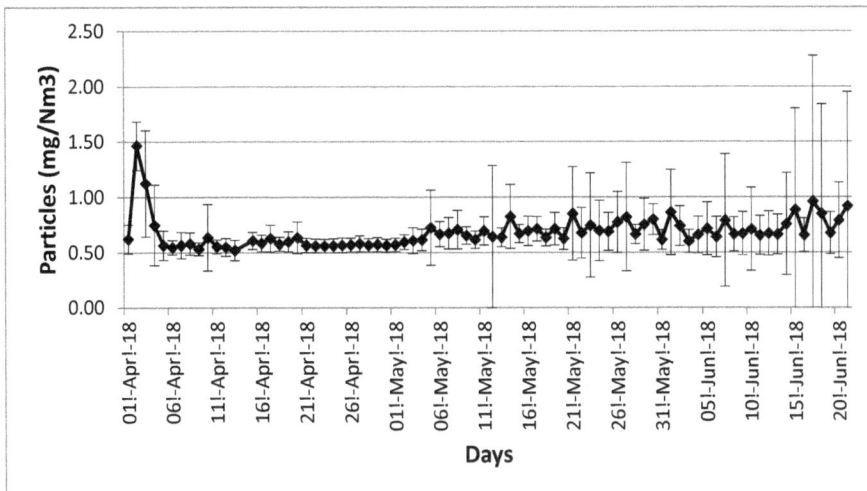

Figure 3. Particle concentration emitted from the stack during the period 1 April 2018 to 21 June 2018.

3.3. Heavy Metals Associated with the Particles

As already stated, the scope of this work was to define the heavy metals retained from the air-pollution control system. As a result, analysis of heavy-metal concentrations in the bag houses, in the dust selected from the tanks of the two gravity settlers, and the two bag houses was performed by using a Wavelength Dispersive X-Ray Fluorescence System (WDXRF, Rigaku, ZSX Primus II) in the laboratory. The Quant Application took place using four national institute of standards and technology NIST standards (NIST 1646a, NIST1468a, NIST 2584, NIST 2710a). Twelve filters were collected from the bag houses: six from the PK720 (furnace bag house), three collected from the first bag in the bag house row, and three from the last bag in the row. From every bag, three samples with a diameter of 40 mm were cut. The first sample was from the upper side of the bag, the second from the middle, and the third one from the bottom. The same procedure was carried out for bag filters from PK820 (sanitary bag house). The dust collected from the tanks of the two gravity settlers and the bag houses was sieved to 60 μm and pressed to a pellet at 1520 kPa. The mean values of the three samples of each bag house filter are shown in Table 1. The results from the dust analysis are shown in Table 2. There are also Supplementary Tables S1 and S2 where LOQ and blanks from bag filters analysis, and LOQ from heavy metal concentrations in the dust from Gravity Settlers and from Bag houses are indiacted, respectively.

Table 1. Heavy metal concentrations in bag houses PK720 (bag house furnace) and PK820 (sanitary bag house) in gr/kg. **N.D.** stands for nondetected. **STDEV** Standard Deviation.

Bag House Filters	PK720 (Bag House Furnace)				PK820 (Sanitary Bag House)			
	Input		Output		Input		Output	
Elements	Average (gr/Kg)	STDEV	Average (gr/Kg)	STDEV	Average (gr/Kg)	STDEV	Average (gr/Kg)	STDEV
Ag	N.D.		N.D.		N.D.		0.05	
Al	0.04	0.00	0.06	0.01	1.55	0.22	0.95	0.09
Bi	N.D.		N.D.		N.D.		N.D.	
Ca	N.D.		0.03	0.02	1.07	0.17	4.68	0.23
Co	N.D.		N.D.		N.D.		N.D.	
Cr	N.D.		N.D.		N.D.		0.06	
Cu	0.17	0.03	0.29	0.09	8.44	2.03	1.23	0.24
Fe	0.02	0.02	0.32	0.17	6.46	0.36	2.28	1.08
Ga	0.68	0.16	0.87	0.11	1.58	0.07	0.43	0.09
K	0.50	0.07	0.60	0.11	N.D.		N.D.	
Mg	N.D.		N.D.		N.D.		N.D.	
Mn	N.D.		N.D.		N.D.		N.D.	
Na	16.90	0.37	22.55	0.82	17.29	0.73	3.29	0.58
Ni	N.D.		N.D.		0.62	0.05	0.13	0.02
Pb	3.80	0.42	4.74	0.15	9.03	0.13	3.53	0.17
Sr	N.D.		N.D.		N.D.		N.D.	
Tl	10.43		N.D.		N.D.		N.D.	
Zn	N.D.		N.D.		N.D.		N.D.	

Table 2. Heavy metal concentrations in the dust from gravity settlers and from bag houses. **N.D.** stands for non-detected.

Elements	Gravity Settlers		Bag Houses	
	MC720 (ppm)	MC820(ppm)	PK720 (ppm)	PK820 (ppm)
Al	N.D.	N.D.	N.D.	N.D.
Ca	N.D.	N.D.	N.D.	N.D.
Fe	990	820	1580	1670
Mg	N.D.	N.D.	N.D.	N.D.
P	90	6680	120	7340
K	N.D.	N.D.	N.D.	N.D.
Si	N.D.	N.D.	N.D.	N.D.
Na	25950	17410	28470	16880
S	N.D.	N.D.	N.D.	N.D.
Ti	N.D.	N.D.	N.D.	N.D.
AS	48	333	172	325
Cd	2501	120	1791	116
Cr	N.D.	N.D.	N.D.	N.D.
Cu	75	395	167	455
Pb	19774	2032	21096	21965
Mn	N.D.	N.D.	N.D.	N.D.
V	N.D.	N.D.	N.D.	N.D.
Zn	172	262	597	222
Ba	229	169	271	302
Ce	N.D.	N.D.	N.D.	N.D.
Co	N.D.	N.D.	N.D.	N.D.
Ga	116	89	105	87
La	N.D.	N.D.	N.D.	N.D.
Mo	4	5	5	4
Nd	63	50	58	48
Ni	14	62	34	75
Rb	20	34	27	35
Sc	N.D.	N.D.	N.D.	N.D.
Sr	42	45	N.D.	48
Th	54	45	51	45
U	N.D.	N.D.	N.D.	N.D.

Table 2. *Cont.*

Elements	Gravity Settlers		Bag Houses	
	MC720 (ppm)	MC820(ppm)	PK720 (ppm)	PK820 (ppm)
Sb	820	3671	1889	3962
Hg	8	7	7	7
Br	421	N.D.	240	N.D.
Cs	N.D.	N.D.	N.D.	N.D.
Bi	26	N.D.	25	21
Sm	N.D.	N.D.	N.D.	N.D.
W	N.D.	N.D.	N.D.	N.D.
Zr	N.D.	N.D.	N.D.	N.D.
Cl	836	641	1142	447

The detailed chemical composition from the filters in Table 1 shows that the highest concentrations in PK720 were for Na, Tl, and Pb. In PK820, large amounts of Cu, Fe, Ca, Na, and Pb were found. PbO and PbO_2 were extracted from the lead-battery scrap, and they reacted with the additive C in the melting furnace to give Pb. The high concentrations of Na came from soda ash (Na_2CO_3), which is used as a flux agent to liquidize the slag. Fe and Cu are additives in the furnace. Fe is used in order to grab sulfur compounds, while Cu is a metal that is used in antimonial alloys for battery construction and is also found in battery poles.

Moreover, there were elements found in PK820 that were absent or in low concentrations in PK720. For example, Ni and Ag was found only in the sanitary bag house and also Ca, Cu, Fe, and Al were in much higher concentrations in PK820 than in PK720. This is because the sanitary air comes from refinery kettles where all these elements are emitted from the refinery process. During the refinery process, the metal impurities (Cu, Zn, Fe, Al, Ni, As, Sn, Sb, Ag, Bi) included in the raw lead must be removed in order to obtain the final lead products: soft lead (which has a high purity, higher than 99.985%), hard lead, and various lead alloys. These metals are removed by performing a series of chemical reactions on the molten lead. A chemical oxidation process is used to remove all residual elements. For example, a mixture of sodium nitrate and sodium hydroxide is added to the molten lead. Alternatively, air enriched with oxygen is added. The mixed oxides produced are then skimmed from the molten lead surface and stored to be fed back to the smelting furnace.

K and Tl come only from the furnace and its surrounding air, and Na is found in much higher concentrations in PK720 filters (furnace) than in PK820 (sanitary air).

There are perceptible differences in heavy-metal concentrations between the first-row filter and the last in every bag house. The concentrations in PK720 were higher in the last-row filter compared to the first for all elements. The opposite was found for PK820. There was no regulation in the way the filter gathers the elements. This probably depends on other parameters, such as the size of the particles trapped in the filter or the way dust removal takes place.

Finally, we calculated the total mass of the Pb emitted in the atmosphere. Provided that we calculated the mean Pb concentration in the bag house filters in gr/kg and used the mean concentration of the dust in the stack, we could find the amount of Pb emitted in the atmosphere. The Pb concentration in the stack was found to be 0.0036 mg Pb/m^3 which is below the upper limit of 0.2 mg/m^3. According to the European Commission, implementing EU Decision 2016/1032, establishing the best available techniques under Directive 2010/75/EU, the limit of Pb in the outflow of the stack should be 0.2 mg/m^3. Calculating this amount with the total air flow passed through the stack during a year, we calculated the total Pb amount emitted in the atmosphere to be almost 3.7 kg.

In Table 2, the element concentrations (ppm) from the dust in the two gravity settler tanks (MC720 and MC820) and from the two bag house tanks (PK720 and PK820) are listed. As expected, chemical analysis highlighted variable and high Na and Pb metal content in the dust obtained from all tanks. According to these results, the dust from the sanitary air mainly contained P, Fe, Na, Pb, and Sb. For the furnace air, the main metals found were Fe, Na, Cd, Pb, Sb, and Cl. Another interesting issue is that a high concentration of Sb was found in the dust from sanitary air. Furthermore, metals including

As and Cu were found mainly in the dust from the sanitary air, and Zn mainly in the dust from the furnace air. Finally, other impurities were found in the samples were Nd, Ni, Rb, Sr, Th, Hg, and Bi.

Ettler et al. [21] conducted a survey and studied the mineralogy and solubility of air pollution control residues from a secondary lead smelter. Comparing the results from our study to this survey we found that the values of heavy metals in bag filters from furnaces are comparable for Fe and Cu. Pb, Ca, Al, Na, and K concentrations in our study were lower, and only Ca showed a higher value. This is probably due to the fact that these two industries follow different methods in production/recovery lines.

There have been several studies conducted to investigate the influence of heavy metals on the environment and also on human-health [25–31]. Early life exposure to heavy metals, such as Cobalt (Co), Copper (Cu), Thalium (Tl), and Selenium (Se) has negative effects to human development [32]. High concentrations of Cadmium (Cd), Copper (Cu), Lead (Pb), and Arsenic (As) increase the health risks to children in contaminated areas [33]. High Pb concentrations have caused regional contaminations around the world [27,34–36]. Health effects from exposure to Cd include problems to kidneys [37] and bones [38]. According to Lanphera et al. [39], even low-level lead exposure increases the risk factor for death, and particularly for cardiovascular-disease death. This is also supported by Obeng-Gyasi et al. [40]. Furthermore, lead could harm the function of kidneys [41,42]. High lead exposure in the work-place has been associated with adverse hepatobiliary clinical makers [43] and with problems in liver function [44].

4. Conclusions

Particle removal efficiency of the whole system was calculated to be 0.9991. It was found that Pb and Na had the highest concentrations in both the PK720 and PK820 filters. Furthermore, Ag and Ni were found only in sanitary bag house filters, with Ca, Cu, Fe, and Al in higher concentrations than in the furnace bag house filters. This is because the particles in the sanitary bag house filters come from chemical reactions performed in metallurgical kettles for secondary lead refinement. Particles from furnace fumes include elements not found in sanitary bag house filters, such as Tl and K, or found in much higher concentrations, such as Na. There are also differences in heavy metals between the first row filters and the last-row ones in every bag house. Finally, it was calculated that the Pb emitted annually in the atmosphere is ca. 4 kg.

Concerning the dust from different areas of the air-pollution control system, chemical analysis highlighted the high metal content of Na and Pb in all cases. However, the highest Na concentrations were found in the dust from MC720 and PK720 (furnace) and, for Pb, the concentration was much lower in the dust from sanitary gravity settler MC820. The dust from the sanitary fumes was mainly enriched with P, Fe, Na, Sb, and Pb, and from the furnace mainly with Fe, Na, Cd, Pb, Sb, and Cl. In all cases, impurities of Nd, Ni, Rb, Sr, Th, Hg, and Bi were found.

This study determined a total range of unintentional heavy-metal emissions from a lead battery recycling process. This information should be useful for understanding the concentrations of heavy metals produced by this process, and in developing a heavy-metal emission inventory. More investigations need to be performed to obtain further information.

The obtained results indicate that the air-pollution control system trapped high amounts of particles containing toxic metals. Measurements for heavy metals, indoors and outdoors, should be performed in the future to confirm that the air-pollution control system could contribute to reducing the potential environmental and health impacts in the area.

Supplementary Materials: The following are available online at http://www.mdpi.com/1996-1073/11/12/3465/s1, Table S1: LOQ and blanks from bag filters analysis values in µg/cm², Table S2: LOQ in ppm from heavy metal concentrations in the dust from Gravity Settlers and from Bag houses.

Author Contributions: Conceptualization, K.K.; Methodology, G.L.; Validation, G.L.; Formal Analysis K.K., D.K. and A.S.; Investigation, K.K., D.K., and A.S.; Resources A.S.; Data curation, K.K., D.K. and A.S.; Writing-Original Draft Preparation, K.K.; Writing-Review & Editing, K.K. and A.S.; Visualization, K.K. and A.S.; Supervision, G.L.

Funding: This research received no external funding.

Acknowledgments: The authors acknowledge the Sunlight recycling factory management team and mainly CEO S. Kopolas for the supply of data in order to conduct our research.

Conflicts of Interest: The authors declare no conflicts of interest.

References

1. Chen, H.Y.; Li, A.J.; Finlow, D.E. The lead and lead-acid battery industries during 2002 and 2007 in China. *J. Power Sources* **2009**, *191*, 22–27. [CrossRef]

2. Gottesfeld, P.; Pokhrel, A.K. Review: Lead Exposure in Battery Manufacturing and Recycling in Developing Countries and Among Children in Nearby Communities. *J. Occup. Environ. Hyg.* **2011**, *8*, 520–532. [CrossRef] [PubMed]

3. Uzu, G.; Sobanska, S.; Sarret, G.; Sauvain, J.J.; Pradère, P.; Dumat, C. Characterization of lead-recycling facility emissions at various workplaces: Major insights for sanitary risks assessment. *J. Hazard. Mater.* **2011**, *186*, 1018–1027. [CrossRef] [PubMed]

4. Tian, X.; Wu, Y.; Gong, Y.; Agyeiwaa, A.; Zuo, T. Residents' behavior, awareness, and willingness to pay for recycling scrap lead-acid battery in Beijing. *J. Mater. Cycles Waste Manag.* **2015**, *17*, 655–664. [CrossRef]

5. Rada, S.; Unguresan, M.L.; Bolundut, L.; Rada, M.; Vermesan, H.; Pica, M.; Culea, E. Structural and electrochemical investigations of the electrodes obtained by recycling of lead acid batteries. *J. Electroanal. Chem.* **2016**, *780*, 187–196. [CrossRef]

6. Pan, J.; Zhang, C.; Sun, Y.; Wang, Z.; Yang, Y. A new process of lead recovery from waste lead-acid batteries by electrolysis of alkaline lead oxide solution. *Electrochem. Commun.* **2012**, *19*, 70–72. [CrossRef]

7. Chen, T.; Zhan, M.-X.; Lin, X.-Q.; Li, Y.-Q.; Zhang, J.; Li, X.-D.; Yan, J.-H.; Buekens, A. Emission and distribution of PCDD/Fs and CBzs from two co-processing RDF cement plants in China. *Environ. Sci. Pollut. Res.* **2016**, *23*, 11845–11854. [CrossRef]

8. Cobo, M.; Gálvez, A.; Conesa, J.A.; de Correa, C.M. Characterization of fly ash from a hazardous waste incinerator in Medellin, Colombia. *J. Hazard. Mater.* **2009**, *168*, 1223–1232. [CrossRef]

9. Zhu, F.; Takaoka, M.; Shiota, K.; Oshita, K.; Kitajima, Y. Chloride Chemical Form in Various Types of Fly Ash. *Environ. Sci. Technol.* **2008**, *42*, 3932–3937. [CrossRef]

10. Korotkova, T.G.; Ksandopulo, S.J.; Bushumov, S.A.; Burlaka, S.D.; Say, Y.V. Quantitative Chemical Analysis of Slag Ash of Novocherkassk State District Power Plant. *Orient. J. Chem.* **2017**, *33*, 186–198. [CrossRef]

11. Liu, G.; Yang, L.; Zhan, J.; Zheng, M.; Li, L.; Jin, R.; Zhao, Y.; Wang, M. Concentrations and patterns of polychlorinated biphenyls at different process stages of cement kilns co-processing waste incinerator fly ash. *Waste Manag.* **2016**, *58*, 280–286. [CrossRef] [PubMed]

12. Nie, Z.; Zheng, M.; Liu, W.; Zhang, B.; Liu, G.; Su, G.; Lv, P.; Xiao, K. Estimation and characterization of PCDD/Fs, dl-PCBs, PCNs, HxCBz and PeCBz emissions from magnesium metallurgy facilities in China. *Chemosphere* **2011**, *85*, 1707–1712. [CrossRef] [PubMed]

13. Nie, Z.; Liu, G.; Liu, W.; Zhang, B.; Zheng, M. Characterization and quantification of unintentional POP emissions from primary and secondary copper metallurgical processes in China. *Atmos. Environ.* **2012**, *15*, 109–115. [CrossRef]

14. Xueli, N.; Henggen, S.; Yinghui, W.; Liuke, Z.; Xingcheng, L.; Min, F. Investigation of the pyrolysis behaviour of hybrid filter media for needle-punched nonwoven bag filters. *Appl. Therm. Eng.* **2017**, *113*, 705–713. [CrossRef]

15. Quina, M.J.; Bordado, J.C.; Quinta-Ferreira, R.M. Treatment and use of air pollution control residues from MSW incineration: An overview. *Waste Manag.* **2008**, *28*, 2097–2121. [CrossRef]

16. Songa, G.-J.; Kima, K.-H.; Seoa, Y.-C.; Kimb, S.-C. Characteristics of ashes from different locations at the MSW incinerator equipped with various air pollution control devices. *Waste Manag.* **2004**, *24*, 99–106. [CrossRef]

17. Ma, Y.; Bai, H.; Zhao, L.; Ma, Y.; Cang, D. Study on the Respirable Particulate Matter Generated from the Petroleum Coke and Coal Mixed-fired CFB Boiler. In Proceedings of the 2010 International Conference on Digital Manufacturing and Automation, Changsha, China, 18–20 December 2010.

18. Ribeiro, J.P.; Vicente, E.D.; Alves, C.; Querol, X.; Amato, F.; Tarelho, L.A.C. Characteristics of ash and particle emissions during bubbling fluidised bed combustion of three types of residual forest biomass. *Environ. Sci. Pollut. Res.* **2017**, *24*, 10018–10029. [CrossRef]

19. Sobanska, S.; Ricq, N.; Laboudigue, A.; Guillermo, R.; Bremard, C.; Laureyns, J.; Merlin, J.C.; Wignacourt, J.P. Microchemical Investigations of Dust Emitted by a Lead Smelter. *Environ. Sci. Technol.* **1999**, *33*, 1334–1339. [CrossRef]

20. Spear, T.M.; Svee, W.; Vincent, J.H.; Stanisich, N. Chemical Speciation of Lead Dust Associated with Primary Lead Smelting. *Environ. Health Perspect.* **1998**, *106*, 565–571. [CrossRef]

21. Ettler, V.; Johan, Z.; Baronnet, A.; Jankovsky, F.; Gilles, C.; Michaljevich, M.; Sebek, O.; Strand, L.; Bezdicka, P. Mineralogy of Air-Pollution-Control Residues from a Secondary Lead Smelter: Environmental Implication. *Environ. Sci. Technol.* **2005**, *39*, 9309–9316. [CrossRef]

22. Liu, J.X.; Chang, D.Q.; Xie, Y.; Mao, N.; Sun, X. Research on fine particles capture of baghouse filter media. *Appl. Mech. Mater.* **2013**, 1293–1297. [CrossRef]

23. Nevers, N.D. *Air Pollution Control Engineering*; Waveland Press: Long Crove, IL, USA, 2000.

24. Rapsomanikis, S.; Kastrinakis, E. *Ai Pollution Control*; Tziolas Publications: Thessaloniki, Greece, 2009.

25. Huang, J.-H.; Ilgen, G.; Matzner, E. Fluxes and budgets of Cd, Zn, Cu, Cr and Ni in a remote forested catchment in Germany. *Biogeochemistry* **2011**, *103*, 59–70. [CrossRef]

26. Spurgeon, D.J.; Lawlor, A.; Hooper, H.L.; Wadsworth, R.; Svendsen, C.; Thomas, L.D.K.; Ellis, J.K.; Bundy, J.G.; Keun, H.C.; Jarup, L. Outdoor and indoor cadmium distributions near an abandoned smelting works and their relations to human exposure. *Environ. Pollut.* **2011**, *159*, 3425–3432. [CrossRef]

27. Miller, E.K.; Friedland, A.J. Lead Migration in Forest Soils: Response to Changing Atmospheric Inputs. *Environ. Sci. Technol.* **1994**, *28*, 662–669. [CrossRef]

28. Gaetke, L.M.; Chow-Johnson, H.S.; Chow, C.K. Copper: Toxicological relevance and mechanisms. *Arch. Toxicol.* **2014**, *88*, 1929–1938. [CrossRef]

29. Jordanova, M.; Hristovski, S.; Musai, M.; Boškovska, V.; Rebok, K.; Dinevska-Ќovkarovska, S.; Melovski, L. Accumulation of Heavy Metals in Some Organs in Barbel and Chub from Crn Drim River in the Republic of Macedonia. *Bull. Environ. Contam. Toxicol.* **2018**, *101*, 392–397. [CrossRef]

30. Mol, S.; Kahraman, A.E.; Ulusoy, S. Potential Health Risks of Heavy Metals to the Turkish and Greek Populations via Consumption of Spiny Dogfish and Thornback Ray from the Sea of Marmara. *Turk. J. Fish. Aquat. Sci.* **2018**, *19*, 109–117. [CrossRef]

31. Merian, E.; Anke, M.; Ihnat, M.; StoeppJer, M. *Elements and Their Compounds in the Environment*; WILEY-VCH Verlag GmbH and Co. KGaA: Weinheim, Germany, 2004.

32. Silver, M.K.; Arain, A.L.; Shao, J.; Chen, M.; Xia, Y.; Lozoff, B.; Meeker, J.D. Distribution and predictors of 20 toxic and essential metals in the umbilical cord blood of Chinese newborns. *Chemosphere* **2018**, *210*, 1167–1175. [CrossRef]

33. Cai, L.-M.; Wang, Q.-S.; Luo, J.; Chen, L.-G.; Zhu, R.-L.; Wang, S.; Tang, C.-H. Heavy metal contamination and health risk assessment for children near a large Cu-smelter in central China. *Sci. Total Environ.* **2019**, *650*, 725–733. [CrossRef]

34. Klaminder, J.; Bindler, R.; Emteryd, O.; Renberg, I. Uptake and recycling of lead by boreal forest plants: Quantitative estimates from a site in northern Sweden. *Geochim. Cosmochim. Acta* **2005**, *69*, 2485–2496. [CrossRef]

35. Klaminder, J.; Bindler, R.; Emteryd, O.; Appleby, P.; Grip, H. Estimating the mean residence time of lead in the organic horizon of boreal forest soils using 210-lead, stable lead and a soil chronosequence. *Biogeochemistry* **2006**, *78*, 31–49. [CrossRef]

36. Zhou, J.; Du, B.; Wang, Z.; Zhang, W.; Xu, L.; Fan, X.; Liu, X.; Zhou, J. Distributions and pools of lead (Pb) in a terrestrial forest ecosystem with highly elevated atmospheric Pb deposition and ecological risks to insects. *Sci. Total Environ.* **2019**, *647*, 932–941. [CrossRef] [PubMed]

37. Thomas, L.D.K.; Hodgson, S.; Nieuwenhuijsen, M.; Jarup, L. Early Kidney Damage in a Population Exposed to Cadmium and Other Heavy Metals. *Environ. Health Perspect.* **2009**, *117*, 181–184. [CrossRef]

38. Nordberg, G.F. Historical perspectives on cadmium toxicology. *Toxicol. Appl. Pharmacol* **2009**, *238*, 192–200. [CrossRef] [PubMed]

39. Lanphear, B.P.; Rauch, S.; Auinger, P.; Allen, R.W.; Hornung, R.W. Low-level lead exposure and mortality in US adults: A population-based cohort study. *Lancet Public Health* **2018**. [CrossRef]

40. Obeng-Gyasi, E.; Armijos, R.X.; Weigel, M.M.; Filippelli, G.M.; Sayegh, M.A. Cardiovascular-Related Outcomes in U.S. Adults Exposed to Lead. *Int. J. Environ. Res. Public Health* **2018**, *15*. [CrossRef]

41. Harari, F.; Sallsten, G.; Christensson, A.; Petkovic, M.; Hedblad, B.; Forsgard, N.; Melander, O.; Nilsson, P.M.; Yan Borne, G.E.; Barregard, L. Blood Lead Levels and Decreased Kidney Function in a Population-Based Cohort. *Am. J. Kidney Dis.* **2018**, in press. [CrossRef]

42. Lin, J.-L.; Lin-Tan, D.-T.; Hsu, K.-H.; Yu, C.-C. Environmental Lead Exposure and Progression of Chronic Renal Diseases in Patients without Diabetes. *N. Engl. J. Med.* **2003**, *348*, 277–286. [CrossRef]

43. Obeng-Gyasi, E.; Armijos, R.X.; Weigel, M.M.; Filippelli, G.; Sayegh, M.A. Hepatobiliary-Related Outcomes in US Adults Exposed to Lead. *Environments* **2018**, *5*. [CrossRef]

44. Can, S.; Bağcı, C.; Ozaslan, M.; Bozkurt, A.; Cengiz, B.; Çakmak, E.A.; Kocabaş, R.; Karadağ, E.; Tarakçıoğlu, M. Occupational lead exposure effect on liver functions and biochemical parameters. *Acta Physiol. Hung.* **2008**, *95*, 395–403. [CrossRef]

energies

MDPI

Article

The Impact of Fuel Type on the Output Parameters of a New Biofuel Burner

Karol Tucki [1,*], Olga Orynycz [2,*], Andrzej Wasiak [2], Antoni Świć [3] and Joanna Wichłacz [1]

[1] Department of Organization and Production Engineering, Warsaw University of Life Sciences, Nowoursynowska Street 164, 02-787 Warsaw, Poland; joannawichlacz@onet.pl

[2] Department of Production Management, Bialystok University of Technology, Wiejska Street 45A, 15-351 Bialystok, Poland; a.wasiak@pb.edu.pl

[3] Faculty of Mechanical Engineering, Institute of Technological Information Systems, Lublin University of Technology, Nadbystrzycka 38 D, 20-618 Lublin, Poland; a.swic@pollub.pl

* Correspondence: karol_tucki@sggw.pl (K.T.); o.orynycz@pb.edu.pl (O.O.); Tel.: +48-746-98-40 (O.O.)

Received: 26 February 2019; Accepted: 8 April 2019; Published: 10 April 2019

Abstract: Intensified action aimed at reducing CO_2 emissions and striving for energy self-sufficiency of both business entities and individual consumers are forcing the sustainable development of environmentally friendly and renewable energy sources. The development of an appropriate class of equipment and production technology is not without significance in this process. On the basis of a proven design for a combustion burner for ecological fuels, a new biofuel burner, also dedicated to prosumers' energetics, was built. The aim of the study was to determine the effect of the type of biofuel on a burner's output parameters, especially gaseous emissions, during the combustion of four types of fuels, including three types of biomass. The combustion temperature was measured for lignite, wood pellets, straw pellets, and sunflower pellets. An analysis of exhaust gas composition was performed for lignite and wood pellets. The results of exhaust emissions and combustion temperatures were compared with the burners currently in use. The use of a new burner might contribute to cleaner combustion and reducing the emissions of some gaseous components.

Keywords: biofuel burner; combustion; ecological fuels; energy management; cleaner combustion

1. Introduction

Declining reserves of fossil fuels [1,2] and a marked increase in their prices combined with the emission of dangerous and harmful carbon dioxide into the atmosphere are just some of the factors that cause interest in renewable energy sources [3,4]. In addition to energy from water, wind, or solar radiation, biomass is the most commonly used source of renewable fuel [5,6]. Its importance as a clean, inexpensive, and renewable source of energy is clear, particularly when it is produced from agricultural and forestry remains [7,8].

The moisture content of biomass has a significant impact on the content of flammable parts in ash and slag [9,10], the combustion efficiency [11,12] and the place of ignition [13,14]. A high content of water vapor in the flue gas reduces the temperature in the furnace, thus reducing the combustion temperature [15,16]. Delivering to the burner a fuel with proper grinding and humidity [17,18], maintaining the appropriate level of air gradation, and even guaranteeing optimal temperature and burning time [19,20] are the parameters guaranteeing correct ignition, complete and total combustion, as well as low emission of pollutants in the exhaust gases [21,22].

Biomass resources can be a fuel for a number of large steam power plants as well as for small production plants [23,24]. Combustion can be carried out using devices and installations designed exclusively for biomass, but also as an auxiliary fuel for coal combustion [25,26]. As in the case of coal, biomass is burned in boilers with a dust or grate furnace [27,28].

Knowledge of the parameters affecting the combustion process positively and unfavorably allows for improvement and proper selection of fuel for a given type of furnace chamber [29,30].

Also, the design of burners for various fuels has become a research priority, searching for optimal flame structure and temperature distribution to assure the most effective combustion [31,32].

Similar strategies are involved in the construction of burners for biofuel combustion. Biomass burning boilers should contain two separate spaces: one for degassing and one for burning a complete part of volatile fuel products [33,34].

The ecological fuel burner consists of a furnace assembly, air preparation and supply unit, fuel tank connected by a screw conveyor with a combustion chamber, and ash gutter with auger assembly of ash extractor (the invention is protected by Polish Patent No. 221180—Boiler for ecological fuels, especially brown coal). The combustion chamber is cylindrical and has left-handed or clockwise openings through which the air is forced through the air preparation and supply unit. In addition, the combustion chamber has, in the upper part, a hot exhaust outlet connected to a heat exchanger. Outside of the combustion chamber there is a hopper, which is connected with a combustion chamber using a screw conveyor. In addition, in the lower part of the combustion chamber there is an ash gutter with blast nozzles, and below the ash gutter there is a screw conveyor of ash extraction. The combustion chamber is partially surrounded by a water cooling unit with heat recovery. The main advantage of the burner is the possibility of using ecological fuel (such as wood chips, briquettes from sawdust, and other fuels), which is rarely used due to the high water content. The pre-drying and drying system that takes place during the transport of fuel to the combustion chamber increases the energy value of the fuel and the vortex motion in the chamber allows for effective combustion. Another advantage is the high combustion efficiency of 94% (in the case of brown coal).

The aim of the study was to determine the effect of the type of biofuel on a burner's output parameters, especially gaseous emissions, during the combustion of four types of fuels, including three types of biomass. The new biofuel burner is compared with the old burner for use with ecological fuels. The combustion of lignite, wood pellet, straw pellets, and sunflower pellets was analyzed, mainly having in mind the flame behavior, temperature distribution, and gaseous emissions.

2. The Design of a New Biofuel Burner

The construction of a new biofuel burner was created on the basis of the invention described above. The idea for a new version of the burner called a new biofuel burner was to reduce the burner's dimensions while maintaining the thermal efficiency of the device.

Differences and similarities between the burner for ecological fuels and the new biofuel burner are summarized in Table 1.

Table 1. Comparison of the new and old designs of burners for ecological fuels.

Construction Element	New Biofuel Burner	Burner for Ecological Fuels
Hopper	100 kg	300 kg
Drying of the fuel	using the heat of combustion	additional fan placed at the hopper
Fuel feeding	screw feeder driven by a motoreducer	screw feeder driven by a motoreducer
Supply of combustion air	through openings in the combustion chamber	through right- or left-handed holes in the combustion chamber
Preparing the combustion air	The fan placed in the upper part of the combustion chamber	The fan placed on the side of the combustion chamber
The outlet of hot exhaust gases	in the axis of the screw feeder	when the holes are left-handed-on the left side; when the holes are clockwise-on the right
Control of the device	control panel	control panel

The new biofuel burner is smaller, simpler in design, and contains fewer components—which reduces the risk of device failure. What distinguishes it from the previous version is a built-in fuel degassing system and ash cooling system protecting against slagging (sinking).

The new biofuel burner is equipped with a fuel drying system that uses warm air from the combustion process. Since the drying is achieved by means of the hot air being the byproduct of a system, it increases the thermodynamic effectiveness of the device. A fan that causes a swirl movement of gases is placed in the upper part of the combustion chamber, where volatile parts of the fuel appear. Similarly to the first solution, the new biofuel burner is equipped with a charging hopper, a screw feeder embedded in the pipe, a cylindrical combustion chamber with a water jacket, and an automatic ash extraction system. The schematic picture of the burner is shown in Figure 1. A biofuel burner, due to its dimensions, can be placed on the trolley to create a mobile system for the production of process steam, in the event of the failure of a regular source of steam in large industrial plants. Also, the small dimensions of the new burner extend the application towards a wider range of consumers with smaller heating power needs.

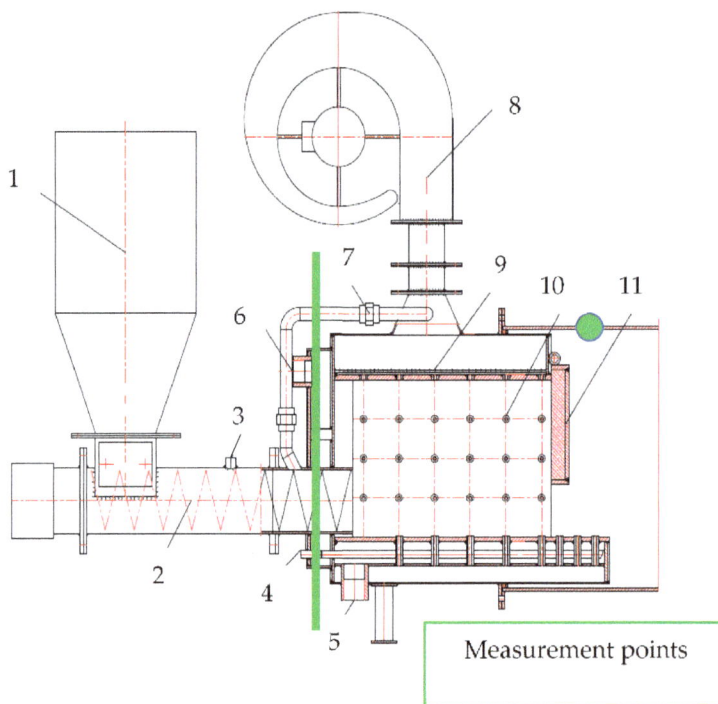

Figure 1. Points for biofuel burner measurements: 1—hopper; 2—screw feeder; 3—fire protection; 4—ash extraction with desludging the burner; 5—burner cooling water inlet; 6—burner cooling water outlet; 7—air supply to the screw; 8—fan; 9—air preparation unit; 10—openings supplying air to the furnace; 11—connection with the boiler.

The fuel can be supplied from the built-in charging hopper or from the storage room adjacent to the boiler room. The fuel is transported by means of a gearmotor (a device consisting of an electric motor and a gear reducer, constituting the drive of the auger, which is connected by means of a collar with a keyway). During transport, the fuel is heated and dried through the air pressed through the pipe into the screw housing. The fuel particles heat up and give off excess water, and further heating triggers the pyrolysis process. Then the fuel is directed to a combustion chamber made of heat-resistant sheet and undergo degassing. After degassing, the volatile parts of the fuel in the vortex motion burn off completely, and the solid parts are pushed into the combustion chamber of the boiler. The burner combustion pressure is approximately 0.024 MPa (0.24 bar). The flue gas under this pressure is directed to the steam generator or steam boiler, heating the water and evaporating the steam. The burner operation is controlled by means of a control cabinet, thanks to which it is possible to control the fuel supply and the amount of air. To facilitate the operation of the burner, a controller system based on the increase of temperature and pressure was used, which enables the system to select the ratio of fuel and air, thereby ensuring the ideal composition of the combustible mixture.

Most of the biomass burning technologies currently used in burners use air vortex movement [35,36]. These constructions have a rotating combustion chamber and blades that cause a swirl of air in the combustion chamber [37–39]. A biofuel burner has a cylindrical, fixed combustion chamber, and the air vortex is forced by the fan at 1000 Pa pressure. A detailed comparison of the biofuel with other solutions available on the Polish market is presented in Table 2.

Table 2. Comparison of burners available on the market.

Parameter	Biofuel Burner	Burner Scanbio [1]	Burner Kipi [2]
Efficiency [%]	94	95	96
Average power consumption [W]	2000	400	600
Fuel consumption [kg/h]	75	67	57
Hopper capacity [kg]	200 kg	No integrated fuel tank—connection required with the warehouse	1800 l
Depth of combustion chamber [mm]	687	700	631
Flue gas temperature (wood pellet) [°C]	1100	220	450

[1] The Scanbio BIOTEC torch is a retort burner controlled by a boiler thermostat. The pellets are delivered to a container inside the device, through a feeder from the fuel storage, then directed to the combustion chamber. The air supply is carried out using a fan. [2] The Kipi rotor burner is a solution based on the use of a rotary combustion chamber and a blower, which eliminates the formation of ashes and gangrene.

3. Materials and Methods

3.1. Materials

The behavior of the following fuels was compared for both the old burner and the new biofuel burner. The fuels use brown coal (lignite), wood pellets, straw pellets, and sunflower seed pellets. The pine pellet, containing an admixture of deciduous trees, was manufactured by King Pellet. The straw pellets made of cereal straw was produced by Polenergia. The physicochemical properties of all fuels used are summarized in Table 3, based on the manufacturer's data.

Table 3. Physicochemical properties of the fuels.

Fuel	Parameter	Value
Lignite [40]	The calorific value	14.882 [MJ/kg]
	Carbon (by mass)	42.2%
	Hydrogen (by mass)	3.17%
	Sulfur (by mass)	1.18%
	Oxygen (by mass)	12.27%
	Nitrogen (by mass)	0.52%
	Moisture (by mass)	31.82%
	Ash (by mass)	8.84%
Wood pellets	The calorific value	19.4 MJ/kg
	Moisture content	7.8% (measurement 10%)
	Ash content	0.47%
	Diameter	6 mm

Table 3. *Cont.*

Fuel	Parameter	Value
Straw pellets	The calorific value	14.218 MJ/kg
	Moisture content	9.3% (measurement 12.6%)
	Ash content	7.6%
	Diameter	6 mm
Sunflower seed pellets	The calorific value	18 MJ/kg
	Moisture content	9% (measurement 10.2%)
	Ash content	0.5%
	Diameter	6 mm

3.2. Methodology

The subject of the research was a comparison of the combustion of several fuels in a burner for ecological fuels and the construction of a new burner for biofuels (biofuel burner).

The following measurements were made on the same types of fuels for both devices:

3.2.1. Humidity of the Fuel

To determine the humidity of the pellets, a pellet moisture meter Tanel PEL-20 (produced by MERA Poland) with a measuring range 10–20% and accuracy of 0.1% was used.

3.2.2. Flue Gas Analysis

The following measurements were made on the same type of fuel for both devices. For the purpose of determining the composition of the flue gas, an exhaust gas analyzer TESTO 320 basic, with a measuring range of 0–4000 ppm and accuracy of 1 ppm was used.

3.2.3. Temperature Measurements

The apparatus used to measure the temperature was a thermovision camera and a pyrometer. The first temperature measurement was made one hour after lighting, the second after one and a half hours, and the third after an hour and 45 min. The accuracy and range of the measuring devices used are given in Table 4.

Table 4. Accuracy and measuring range of devices.

Device	Measuring Range	Accuracy
Thermal imaging camera	−20–1500 °C	0.1 °C
Video Pyrometer AX 7550	−50–1600 °C	0.1 °C

4. Results and Discussion

Temperature measurements were carried out at the following stages and locations in the burner:

- drying of fuel,
- fuel combustion,
- secondary air,
- inside the combustion chamber 100 mm from the axis of the burner,
- inside the combustion chamber 180 mm from the axis of the burner,
- flue gas outlet to the boiler.

The points of measurements are shown in Figure 1.

The temperature distribution profile across the burner was also determined. A graph showing the temperature distribution in relation to the diameter of the biofuel burner is presented in Figure 2.

It is seen that the temperature inside the combustion chamber is distributed symmetrically. The highest temperatures inside the combustion chamber were noted at the edges of the walls, then, approaching the axis of the screw feeder, they decreased; whereas at the same fuel inlet from the feeder to the chamber a temperature rise was again noted. The laboratory combustion curve from the literature coincides with the results of the measurements carried out on the burner, except for lignite (Figure 2). The results of the measurements of the fuels are presented in Table 5.

Table 5. Temperature of combustion for the fuels studied.

Measurement Point	Measurement Number	Temperature of the Lignite Combustion [°C]	Temperature of the Wood Pellet Combustion [°C]	Temperature of Straw Pellets Burning [°C]	Temperature of the Sunflower Pellet Burning [°C]
The air pipe for drying the fuel	I	480	600	440	580
	II	500	621	448	587
	III	502	615	452	584
Fuel inlet to the combustion chamber	I	322	545	360	539
	II	350	604	378	542
	III	348	608	375	542
External wall of the combustion chamber	I	978	1167	904	1087
	II	991	1142	892	1111
	III	1000	1169	897	1100
100 mm distance from the screw	I	408	408	256	370
	II	433	433	249	371
	III	429	429	251	370
180 mm distance from the screw	I	760	760	694	854
	II	799	799	690	859
	III	802	802	694	862
Exhaust gas outlet	I	1064	1209	1002	1201
	II	1103	1250	987	1215
	III	1100	1244	990	1218

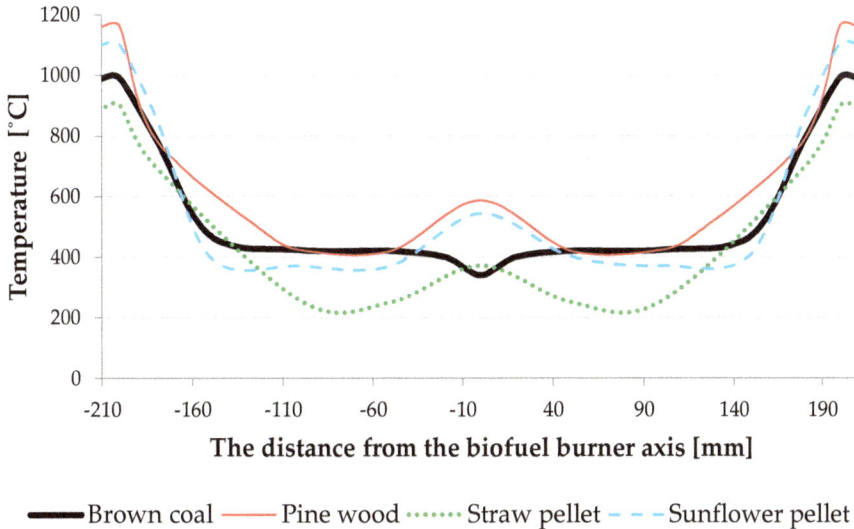

Figure 2. Comparison of the results of combustion of various types of fuels in the biofuel burner.

4.1. The Exhaust Gas Analysis

To determine the emission level of the burner (in the case of brown coal) and check whether it meets the emission standards regarding the content of impurities in the flue, the gas analysis was compared to the requirements of PN-EN 303-5/2012. The results are summarized in Table 6.

Table 6. Results of measurements from the exhaust gas analyzer.

Exhaust Gas Component	Measurement	Requirement of the Standard
O_2	7.7%	met
CO_2	9.8%	met
CO	68 ppm 116 mg/kWh	met
NO	5 ppm 14%CO_2	met
NO_x	12 g/kWh	met

4.2. Summary of Biofuel Burner Characteristics

To determine the total power of the steam generator equipped with the new biofuel burner, it was necessary to calculate the power of the steam boiler, the power of the cooling system of the burner, and the chemical energy of the fuel burned.

In the case of characteristic values of exhaust gases, calculations were made only for lignite, because of all the fuels used for testing, only in the case of brown coal are the fuel composition reports generally available. The composition of lignite is presented in Table 3. The results of all calculations are presented in Table 7.

Table 7. Summary of calculation results.

Parameter	Value	Unit
The power of steam boiler	232	kW
The power of the cooling system	20.32	kW
Chemical energy of the fuel	310	kW
The total power of the system	292	kW
System efficiency	94	%
The minimum oxygen mass demand for burning	1.268	kg O_2/kg fuel
Minimum oxygen molar demand for combustion	0.03963	kmol O_2/kg fuel
Minimum combustion air requirement	0.1887	kmol air/kg fuel
Actual air demand for combustion	0.302	kmol air/kg fuel
Total amount of damped moist exhaust gas	9.6781	kg moist exhaust gas/kg fuel (without ballast)
Composition of moist exhaust gas: CO_2	0.1599	kg CO_2/kg moist exhaust gas
Composition of moist exhaust gas: SO_2	0.0025	kg SO_2/kg moist exhaust gas
Composition of moist exhaust gas: N_2	0.6909	kg N_2/kg moist exhaust gas
Composition of moist exhaust gas: O_2	0.786	kg O_2/kg moist exhaust gas
Composition of moist exhaust gas: H_2O	0.0681	kg H_2O/kg moist exhaust gas
Flux of part of fuel without ash-ballast	0.019	kg/s
Mass flux of moist flue gas flow	0.18	kg/s
Emission CO_2	103.63	kg/h
Emission SO_2	1.62	kg/h
Emission N_2	447.7	kg/h
Emission O_2	50.93	kg/h
Emission H_2O	44.13	kg/h
Emission of dust	0.0663	kg/h

It can be concluded from Table 7 that, e.g., the ash content in flue gas is as low as 0.01% (mass), which appears rather low as compared with the data for classical burners [41,42].

The latter source [42] also reports higher CO_2 content in exhaust gas (11.6%, while our biofuel-burner has 9.8%). It can be concluded, therefore, that higher air excess is assured by the controlling unit, which also results in better combustion.

The brown coal combustion temperature was 1000 °C. Lignite is one of the fuels containing the highest water concentration. When air is injected into the tube with an embedded worm screw, the fuel starts to evaporate and the drying temperature is reduced from 500 °C to 340 °C. At the outlet from the burner to the boiler, the flue gas reached a temperature of about 1100 °C. Ash and moisture reduce the heating value of lignite and, consequently, the output parameters of its combustion.

During the combustion of wood pellets made of sawdust from pine wood, the temperature was read at 1160 °C; at the inlet of the pellet to the furnace the temperature reached 585 °C, and the secondary air temperature was 612 °C. In terms of temperature and emission performance, it is the recommended fuel for this type of equipment.

Straw as biomass for combustion is an attractive fuel due to the low price and high availability. The main problem when operating straw-fired machines is the high silicon content. During harvesting and storage, a large amount of sand accumulates, which during combustion leads to sintering, resulting in a shortening of the life of the burner. In addition, ash from incombustible straw substances has a low melting point of about 95–105 °C, which also results in slagging of the furnace. The combustion temperature compared to other fuels is lower, at 993 °C, and the drying temperature during transport is 371 °C. It can be concluded, therefore, that straw is not an appropriate fuel for the constructed burner.

Pellets from the sunflower husks used for measurements were made with an admixture of sawdust from deciduous trees and oak wood, which increases the calorific value and reduces the susceptibility to sintering. The ash obtained from combustion can be used as an organic fertilizer. The fuel inlet temperature for the combustion chamber is 541 °C, 1100 °C combustion, and the exhaust outlet 1211 °C.

Calculations of exhaust emissions were made using lignite data as it is the only fuel with an elementary chemical composition report. As a result of combustion in the biofuel, there is emission of moist exhausts, i.e., those in which there is water in the form of water vapor. Gases like N_2, O_2, and H_2O are neutral for the environment because they are present in the air. The CO_2 emissions were determined to be 103.62 kg/h, while those of sulfur dioxide (SO_2) were 1.62 kg/h.

The main design strategy of the new biomass burner design was to reduce its size while maintaining combustion efficiency. Analyzing the measurements made and the temperature values read from the burner controller for ecological fuels (Table 8), it can be concluded that the assumption has been met.

Table 8. Comparison of the combustion temperature of the new and old burner designs with ecological fuels.

Fuel Type	Biofuel Burner—Combustion Temperature [°C] *	Ecological Fuel Burner—Combustion Temperature [°C] **
Brown coal	1089.00	1100
Wood pellets	1234.33	1250
Straw pellets	993.00	1000
Sunflower pellets	1211.33	1200

* Based on our own measurements. ** Based on readings from the control device.

The correctness of the results of the conducted tests was also verified by comparison to the values of fuel combustion in a swirl burner presented in [43], considered as a reference.

The fuel whose combustion guarantees very good burning parameters and favorable economic and ecological estimates is fine brown coal [44,45]. The calorific value of granulate 0–20 mm is 9–15 MJ/kg, and the combustion temperature is around 1100 °C [46]. The ash content is only 4–9%, and the sulfur content is 0.42 ± 0.90% (the sulfur content in bituminous coal is 5–12% and is a factor causing corrosion and shortening the life of the burner and the boiler) [47,48]. For comparison, the calorific value of lignite pea coal (granulation: 20–30 mm) is 10–15 MJ, and that of brown coal (granulation: 40–250 mm) is 10–20 MJ. High-calorific brown coal fine imported from the Czech Republic is also available on the Polish market (granulation 0–10 mm, average calorific value 16.9 kJ/kg, average ash content 13.1%, average sulfur content 0.84%, average humidity 29.7%) [49].

The results of the present investigation show that this classic fuel can be replaced by several types of biofuels when an appropriate burner design is used. For comparison, wood pellets and briquettes are burned at a temperature of about 1250 °C, with a calorific value about 19 MJ/kg. Wood chips and energetic willow have a lower thermal efficiency due to the increased humidity. Their combustion temperature is within the range 900–1000 °C. Cereals and straw briquettes reach temperatures up to 1100 °C with a calorific value of 8–14 MJ/kg. Pellets from sunflower husks reach combustion temperature at 1200 °C, and their calorific value is 17–19 MJ/kg [50]. Ash as a secondary product can be used for soil fertilization in agriculture and horticulture [51,52].

5. Conclusions

In the above study, the object of the research was the comparison of the performance of several fuels in a burner for ecological fuels and a new biofuel burner. Based on the conducted analyses, it can be stated that:

- The fuel that achieved the best burning parameters was wood pellets made of pine. Its calorific value is 19.4 MJ/kg and the combustion temperature was 1235 °C.
- In the case of lignite, the average exhaust gas temperature was 1089 °C; for straw pellets it was 993 °C, and for sunflower pellets 1212 °C.
- The measurements made with the exhaust gas analyzer show that the proposed burner solution meets the emission standards regarding the content of impurities in the exhaust.
- The laboratory combustion curve from the literature coincides with the results of the measurements carried out.
- It should be noted that especially emission of ash from a modernized burner is lower than from a traditional one.

Due to the wide application possibilities of the proposed biofuel burner construction, its installation in industrial plants can be an effective pro-ecological solution through cleaner combustion and the reduction of harmful emissions.

It should be emphasized that the problem of air pollution in heating not only applies to the fuel used; the technique and appropriate combustion technology are also important. The wider introduction of such burners, even on the local scale, will ultimately aid the global struggle for a cleaner atmosphere.

Author Contributions: Conceptualization, K.T. and J.W.; Methodology, O.O. and A.W.; Validation, A.Ś. and K.T.; Investigation, O.O. and A.W.; Writing—Original Draft Preparation, A.Ś. and J.W.; Funding Acquisition, A.Ś.

Funding: The authors wish to express gratitude to Lublin University of Technology for financial support given to the present publication (Antoni Świć). The research was carried out under financial support obtained from the research subsidy of the Faculty of Engineering Management (WIZ) of Bialystok University of Technology (Olga Orynycz, Andrzej Wasiak).

Acknowledgments: The authors wish to express their deep gratitude to Lublin University of Technology for the financial support given to the present publication (A.Ś.). The research was carried out with financial support obtained from the research subsidy of the Faculty of Engineering Management (WIZ) of Bialystok University of Technology (O.O., A.W.).

Conflicts of Interest: The authors declare no conflict of interest. The funders had no role in the design of the study; in the collection, analyses, or interpretation of data; in the writing of the manuscript, and in the decision to publish the results.

References

1. Handayani, K.; Krozer, Y.; Filatova, T. From fossil fuels to renewables: An analysis of long-term scenarios considering technological learning. *Energy Policy* **2019**, *127*, 134–146. [CrossRef]
2. Martins, F.; Felgueiras, C.; Smitková, M. Fossil fuel energy consumption in European countries. *Energy Procedia* **2018**, *153*, 107–111. [CrossRef]

3. Mączyńska, J.; Krzywonos, M.; Kupczyk, A.; Tucki, K.; Sikora, M.; Pińkowska, H.; Bączyk, A.; Wielewska, I. Production and use of biofuels for transport in Poland and Brazil—The case of bioethanol. *Fuel* **2019**, *241*, 989–996. [CrossRef]

4. Gökgöz, F.; Güvercin, M.T. Energy security and renewable energy efficiency in EU. *Renew. Sustain. Energy Rev.* **2018**, *96*, 226–239. [CrossRef]

5. Rybak, W. *Spalanie i Współspalanie Biopaliw Stałych*, 1st ed.; Oficyna Wydawnicza Politechniki Wrocławskiej: Wrocław, Poland, 2006; p. 80.

6. Ashter, S.A. Biomass and its sources. In *Technology and Applications of Polymers Derived from Biomass*, 1st ed.; William Andrew: Cambridge, MA, USA, 2018; pp. 11–36.

7. Bunn, D.W.; Redondo-Martin, J.; Munoz-Hernandez, J.I.; Diaz-Cachinero, P. Analysis of coal conversion to biomass as a transitional technology. *Renew. Energy* **2019**, *132*, 752–760. [CrossRef]

8. Hamelin, L.; Borzęcka, M.; Kozak, M.; Pudełko, R. A spatial approach to bioeconomy: Quantifying the residual biomass potential in the EU-27. *Renew. Sustain. Energy Rev.* **2019**, *100*, 127–142. [CrossRef]

9. Royo, J.; Canalis, P.; Quintana, D.; Diaz-Ramirez, M.; Sin, A.; Rezeau, A. Experimental study on the ash behaviour in combustion of pelletized residual agricultural biomass. *Fuel* **2019**, *239*, 991–1000. [CrossRef]

10. Zeng, T.; Weller, N.; Pollex, A.; Lenz, V. Blended biomass pellets as fuel for small scale combustion appliances: Influence on gaseous and total particulate matter emissions and applicability of fuel indices. *Fuel* **2016**, *184*, 689–700. [CrossRef]

11. Lu, J.; Fu, L.; Li, X.; Eddings, E. Capture efficiency of coal/biomass co-combustion ash in an electrostatic field. *Particuology* **2018**, *40*, 80–87. [CrossRef]

12. Kažimírová, V.; Opáth, R. Biomass combustion emissions. *Res. Agric. Eng.* **2016**, *62*, 61–65. [CrossRef]

13. Xu, X.L.; Chen, H.H. Examining the efficiency of biomass energy: Evidence from the Chinese recycling industry. *Energy Policy* **2018**, *119*, 77–86. [CrossRef]

14. Neuenschwander, P.; Good, J.; Nussbaumer, T. Combustion Efficiency in Biomass Furnaces with Flue Gas Condensation. Biomass for Energy and Industry, 10th European Conference and Technology Exhibition. Available online: http://www.bfe.admin.ch/php/modules/enet/streamfile.php?file=000000000374.pdf (accessed on 17 February 2019).

15. Gil, M.V.; González-Vázquez, M.P.; García, R.; Rubiera, F.; Pevida, C. Assessing the influence of biomass properties on the gasification process using multivariate data analysis. *Energy Convers. Manag.* **2019**, *184*, 649–660. [CrossRef]

16. Cheng, J.; Zhou, F.; Si, T.; Zhou, J.; Cen, K. Mechanical strength and combustion properties of biomass pellets prepared with coal tar residue as a binder. *Fuel Process. Technol.* **2018**, *179*, 229–237. [CrossRef]

17. Jamradloedluk, J.; Lertsatitthanakorn, C. Influences of Mixing Ratios and Binder Types on Properties of Biomass Pellets. *Energy Procedia* **2017**, *138*, 1147–1152. [CrossRef]

18. Singh, K.; Zondlo, J. Characterization of fuel properties for coal and torrefied biomass mixtures. *J. Energy Inst.* **2017**, *90*, 505–512. [CrossRef]

19. Xinfeng, W.; Rongrong, G.; Liwei, W.; Wenxue, X.; Yating, Z.; Bing, C.; Weijun, L.; Likun, X.; Jianmin, C.; Wenxing, W. Emissions of fine particulate nitrated phenols from the burning of five common types of biomass. *Environ. Pollut.* **2017**, *230*, 405–412.

20. Xingru, L.; Lei, J.; Yu, B.; Yang, Y.; Shuiqiao, L.; Xi, C.; Jing, X.; Yusi, L.; Yingfeng, W.; Xueqing, G.; et al. Wintertime aerosol chemistry in Beijing during haze period: Significant contribution from secondary formation and biomass burning emission. *Atmos. Res.* **2019**, *218*, 25–33.

21. Guo, F.; Zhong, Z. Optimization of the co-combustion of coal and composite biomass pellets. *J. Clean. Prod.* **2018**, *185*, 399–407. [CrossRef]

22. Xu, G.; Li, M.; Lu, P. Experimental investigation on flow properties of different biomass and torrefied biomass powders. *Biomass Bioenergy* **2019**, *122*, 63–75. [CrossRef]

23. Ahmad, A.A.; Zawawi, N.A.; Kasim, F.H.; Inayat, A.; Khasri, A. Assessing the gasification performance of biomass: A review on biomass gasification process conditions, optimization and economic evaluation. *Renew. Sustain. Energy Rev.* **2016**, *53*, 1333–1347. [CrossRef]

24. Kosowski, K.; Tucki, K.; Piwowarski, M.; Stępień, R.; Orynycz, O.; Włodarski, W.; Bączyk, A. Thermodynamic Cycle Concepts for High-Efficiency Power Plans. Part A: Public Power Plants 60+. *Sustainability* **2019**, *11*, 554. [CrossRef]

25. Kosowski, K.; Tucki, K.; Piwowarski, M.; Stępień, R.; Orynycz, O.; Włodarski, W. Thermodynamic cycle concepts for high-efficiency power plants. Part B: Prosumer and distributed power industry. *Sustainability* **2019**, in press.

26. Stam, A.F.; Brem, G. Fouling in coal-fired boilers: Biomass co-firing, full conversion and use of additives—A thermodynamic approach. *Fuel* **2019**, *239*, 1274–1283. [CrossRef]

27. Kalina, J.; Świerzewski, M.; Strzałka, R. Operational experiences of municipal heating plants with biomass-fired ORC cogeneration units. *Energy Convers. Manag.* **2019**, *181*, 544–561. [CrossRef]

28. Junga, R.; Pospolita, J.; Niemiec, P.; Dudek, M. The assessment of the fuel additive impact on moving grate boiler efficiency. *J. Energy Inst.* **2018**, in press. [CrossRef]

29. Vakkilainen, E.K. Fluidized Bed Boilers for Biomass. In *Steam Generation from Biomass. Construction and Design of Large Boilers*, 1st ed.; Butterworth-Heinemann: Oxford, UK, 2017; pp. 18–179.

30. Zhang, Y.; Li, Q.; Zhou, H. Heat Transfer Calculation in Furnaces. In *Theory and Calculation of Heat Transfer in Furnaces*, 1st ed.; Academic Press: Cambridge, MA, USA, 2016; pp. 131–203.

31. Tinajero, J.; Dunn-Rankin, D. Non-premixed axisymmetric flames driven by ion currents. *Combust. Flame* **2019**, *199*, 365–376. [CrossRef]

32. Weinberg, F.; Carleton, F.; Dunn-Rankin, D. Electric field-controlled mesoscale burners. *Combust. Flame* **2008**, *152*, 186–193. [CrossRef]

33. Kosowski, K. *Steam and Gas Turbines with the Examples of Alstom Technology*; Alstom: Saint-Ouen, France, 2007; ISBN 978-83-925959-3-9.

34. Růžičková, J.; Kucbel, M.; Raclavská, H.; Švédová, B.; Raclavský, K.; Juchelková, D. Comparison of organic compounds in char and soot from the combustion of biomass in boilers of various emission classes. *J. Environ. Manag.* **2019**, *236*, 769–783. [CrossRef]

35. Caposciutti, G.; Barontini, F.; Antonelli, M.; Galletti, C.; Tognotti, L.; Desideri, U. Biomass early stage combustion in a small size boiler: Experimental and numerical analysis. *Energy Procedia* **2018**, *148*, 1159–1166. [CrossRef]

36. Golec, T.; Remiszewski, K.; Świątkowski, B.; Błesznowski, M. Pulverized biomass burners. *Energetyka* **2007**, *5*, 375–382.

37. González-Cencerrado, A.; Peña, B.; Gil, A. Experimental analysis of biomass co-firing flames in a pulverized fuel swirl burner using a CCD based visualization system. *Fuel Process. Technol.* **2015**, *130*, 299–310. [CrossRef]

38. Peña, B.; Pallarés, J.; Bartolomé, C.; Herce, C. Experimental study on the effects of co-firing coal mine waste residues with coal in PF swirl burners. *Energy* **2018**, *157*, 45–53. [CrossRef]

39. Ti, S.; Chen, Z.; Kuang, M.; Li, Z.; Zhu, Q.; Zhang, H.; Wang, Z.; Xu, G. Numerical simulation of the combustion characteristics and NOx emission of a swirl burner: Influence of the structure of the burner outlet. *Appl. Therm. Eng.* **2016**, *104*, 565–576. [CrossRef]

40. Kołodziejczyk, U. Characteristics of brown coal deposits in lubuskie province. *Zeszyty Naukowe. Inżynieria Środowiska/Uniwersytet Zielonogórski* **2010**, *137*, 169–179.

41. Gil, P.; Wilk, J.; Tychanicz, M.; Wielgos, S. Preliminary experimental investigation of the automatic pellet boiler according to the PN-EN 303-5:2012 standard requirements. *Rynek Energii* **2017**, *5*, 74–79.

42. Research Report No 191/18-LG Test of Boiler Type BIO-MAX PELLET 14 Fired with Wood Pellets. Available online: https://www.google.com/url?sa=t&rct=j&q=&esrc=s&source=web&cd=1&cad=rja&uact=8& ved=2ahUKEwiZ57rGmM3gAhVmpYsKHR0hA_wQFjAAegQIAhAC&url=https%3A%2F%2Fpowietrze. malopolska.pl%2Fwp-content%2Fuploads%2F2018%2F12%2FSprawozdanie-BIO-MAX-PELLET-14kW-Pellet-drzewny-.pdf&usg=AOvVaw35t_x_4oYktMpMhr-sZViM (accessed on 22 February 2019).

43. Chaoyang, Z.; Yongqiang, W.; Qiye, J.; Qijuan, C.; Yuegui, Z. Mechanism analysis on the pulverized coal combustion flame stability NOx emission in a swirl burner with deep air staging. *J. Energy Inst.* **2019**, *92*, 298–310.

44. Kasztelewicz, Z.; Tajduś, A.; Cała, M.; Ptak, M.; Sikora, M. Strategic conditions for the future of brown coal mining in Poland. *Energy Policy J.* **2018**, *21*, 155–178.

45. Gilewska, M.; Otremba, K. The some aspects of agricultural reclamation the post-mining grounds of the Konin and Adamów Brown Coal Mines. *Ecol. Eng.* **2018**, *19*, 22–29.

46. Characteristics, Types and Calorific Value of Fuel and Coal. Available online: https://taniepalenie.pl/charakterystyka-rodzaje-kalorycznosc/ (accessed on 20 March 2019).

47. Sieniawa Brown Coal. Available online: http://www.sieniawa.com/en/our-offer/brown-coal/ (accessed on 20 March 2019).

48. InterFuel. Available online: https://www.interfuel.pl/wegiel-brunatny/7 (accessed on 20 March 2019).

49. Czech Brown Coal. Available online: https://czeskiwegiel.wordpress.com/czeski-wegiel-brunatny/ (accessed on 20 March 2019).

50. Igliński, B.; Buczkowski, R.; Cichosz, M. *Technologie Bioenergetyczne*, 1st ed.; Wydawnictwo Naukowe Uniwersytetu Mikołaja Kopernika: Toruń, Polska, 2009; pp. 17–318.

51. Czaja, P.; Kwaśniewski, K. Polish Coal, Energy and Environment—Chances and Dangers. *Annu. Set Environ. Protect.* **2016**, *18*, 38–60.

52. Bożym, M. Fly ash from brown coal in sewage sludge management. *Prace Instytutu Szkła, Ceramiki Materiałów Budowlanych* **2010**, *5*, 104–112.

energies

MDPI

Article

Numerical Investigation of the Effects of Steam Mole Fraction and the Inlet Velocity of Reforming Reactants on an Industrial-Scale Steam Methane Reformer

Chun-Lang Yeh

Department of Aeronautical Engineering, National Formosa University, Huwei, Yunlin 632, Taiwan;
clyeh@nfu.edu.tw; Tel.: +886-5-6315527

Received: 1 July 2018; Accepted: 8 August 2018; Published: 10 August 2018

Abstract: Steam methane reforming (SMR) is the most common commercial method of industrial hydrogen production. Control of the catalyst tube temperature is a fundamental demand of the reformer design because the tube temperature must be maintained within a range that the catalysts have high activity and the tube has minor damage. In this paper, the transport and chemical reaction in an industrial-scale steam methane reformer are simulated using computational fluid dynamics (CFD). Two factors influencing the reformer temperature, hydrogen yield and stress distribution are discussed: (1) the mole fraction of steam (Y_{H2O}) and (2) the inlet velocity of the reforming reactants. The purpose of this paper is to get a better understanding of the flow and thermal development in a reformer and thus, to make it possible to improve the performance and lifetime of a steam reformer. It is found that the lowest temperature at the reforming tube surface occurs when Y_{H2O} is 0.5. Hydrogen yield has the highest value when Y_{H2O} is 0.5. The wall shear stress at the reforming tube surface is higher at a higher Y_{H2O}. The surface temperature of a reforming tube increases with the inlet velocity of the reforming reactants. Finally, the wall shear stress at the reforming tube surface increases with the inlet velocity of the reforming reactants.

Keywords: steam methane reformer; computational fluid dynamics; tube surface temperature; hydrogen yield; wall shear stress

1. Introduction

The environmental impact of greenhouse gas pollutants emitted from the combustion of fossil fuels and the legal regulations against production of air pollutants have increased the necessity for clean combustion. Hydrogen is an efficient energy carrier and is one of the cleanest fuels which can replace fossil fuels. Hydrogen is also an important material for petroleum refineries. It converts crude oil into products with high economic value, e.g., gasoline, jet fuel and diesel. The demand for hydrogen by petroleum refineries has increased due to environmental restrictions and efforts to process heavier components of crude oil. In particular, the environmental requirement for low-sulfur-content fuels results in an increasing amount of hydrogen required in hydro-treating processes, and the attempt to process heavier components of crude oil, known as bottom-of-the-barrel processing, also increases the demand for hydrogen in hydro-cracking processes [1–3].

A wide and generalized use of hydrogen would be feasible only when efficient means of producing it with low emission of pollutants were industrially developed. Hydrogen can be produced by a number of ways, e.g., electrolysis, steam methane reforming (SMR), partial oxidation reforming, nuclear energy, etc. [4,5]. Among these ways, SMR is the most common commercial method of industrial hydrogen production. SMR reaction mainly includes the following three chemical equations:

$$CH_4 + H_2O \rightleftarrows CO + 3H_2 \qquad (1)$$

$$CO + H_2O \rightleftarrows CO_2 + H_2 \qquad (2)$$

$$CH_4 + 2H_2O \rightleftarrows CO_2 + 4H_2 \qquad (3)$$

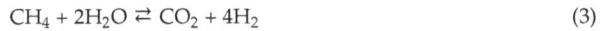

The first (SMR) and the third reactions are endothermic, the second reaction (water-gas shift (WGS)) is exothermic, and the overall reaction is endothermic.

A steam reformer of a hydrogen plant is a device that supplies heat to convert the natural gas or liquid petroleum gas into hydrogen via catalysts. The combustion process in a reformer provides heat to maintain the reforming reaction in a catalyst tube. Control of the catalyst tube temperature is a fundamental demand of the reformer design because the tube temperature must be maintained within a range that the catalysts have high activity and the tube has minor damage. When a steam reformer is operating, the catalyst tubes are subjected to stresses close to the ultimate stress of the tube material. This leads to an acceleration of the creep damage. Safety, reliability and efficiency are the basic requirements of the reformer operation. The catalyst tube should have uniform heat distribution to extend tube life. However, the heat distribution in a reformer is practically non-uniform. In addition, maldistribution of the flue gas and fuel gas may result in flame impingement on the catalyst tubes and lead to localized hot spots and high tube wall temperature. These factors may shorten the tube life.

Owing to the rapid development in computer science and technology, as well as the improvements in physical models and numerical methods, computational fluid dynamics (CFD) is widely used in analyzing systems involving heat transfer, fluid flow and chemical reactions. CFD is also used to simulate systems that cannot be measured easily or simulated experimentally. In recent years, there have been a lot of SMR researches using CFD. Tran et al. [1] developed a CFD model of an industrial-scale steam methane reformer. The authors pointed out that the reformer CFD model can be considered an adequate representation of the on-line reformer and can be used to determine the risk of operating the online reformer at unexplored and potentially more beneficial operating conditions. Di Carlo et al. [2] investigated numerically a pilot scale bubbling fluidized bed SE-SMR (Sorption Enhanced Steam Methane Reforming) reactor by means of a two-dimensional CFD approach. The numerical results show quantitatively the positive influence of carbon dioxide sorption on the reforming process at different operating conditions, specifically the enhancement of hydrogen yield and reduction of methane residual concentration in the reactor outlet stream. Lao et al. [6] developed a CFD model of an industrial-scale reforming tube using ANSYS Fluent with realistic geometry characteristics to simulate the transport and chemical reaction phenomena with approximate representation of the catalyst packing. The authors analyzed the real-time regulation of the hydrogen production by choosing the outer wall temperature profile of the reforming tube and the hydrogen concentration at the outlet as the manipulated input and controlled output, respectively. Mokheimer et al. [7] presented modeling and simulations of SMR process. The model was applied to study the effect of different operating parameters on the steam and methane conversion. The results showed that increasing the conversion thermodynamic limits with the decrease of the pressure results in a need for long reformers so as to achieve the associated fuel reforming thermodynamics limit. It is also shown that not only increasing the steam to methane molar ratio is favorable for higher methane conversion but the way the ratio is changed also matters to a considerable extent. Ni [8] developed a 2D heat and mass transfer model to investigate the fundamental transport phenomenon and chemical reaction kinetics in a compact reformer for hydrogen production by SMR. Parametric simulations were performed to examine the effects of permeability, gas velocity, temperature, and rate of heat supply on the reformer performance. It was found that the reaction rates of SMR and WGS are the highest at the inlet but decrease significantly along the reformer. Increasing the operating temperature raises the reaction rates at the inlet but shows very small influence downstream. Ebrahimi et al. [9] applied a three-dimensional zone method to an industrial fired heater of SMR reactor. The effect of emissivity, extinction coefficient, heat release pattern and flame angle on performance of the fired heater are presented. It was found that decreasing the extinction coefficients of combustion gases by 25% caused 2.6% rise in the temperature of heat sink surfaces. Seo et al. [10] investigated numerically a compact SMR system integrated with a WGS reactor. Heat transfer to the catalyst beds and the catalytic reactions in the SMR and WGS catalyst beds were investigated. The effects of the cooling heat flux at the outside wall of the system and steam-to-carbon ratio were also examined. It was found that as the cooling heat flux increases, both the methane conversion and carbon monoxide content are reduced in the SMR bed and the carbon monoxide conversion is improved in the WGS bed. In addition, both methane conversion and carbon dioxide reduction increase with increasing steam-to-carbon ratio.

In this paper, the transport and chemical reaction in an industrial-scale steam methane reformer are simulated using CFD. Two factors influencing the reformer temperature, hydrogen yield and stress are discussed. They include the mole fraction of steam and the inlet velocity of the reforming reactants. The objective is to get a better understanding of the flow and thermal development in a reformer and thus to make it possible to improve the performance and lifetime of a steam reformer.

2. Numerical Methods and Physical Models

In this study, the ANSYS FLUENT V.17 commercial code [11] is employed to simulate the reacting and fluid flow in a steam methane reformer. The SIMPLE algorithm by Patankar [12] is used to solve the governing equations. The discretizations of convection terms and diffusion terms are carried out by the second order upwind scheme and the central difference scheme, respectively. In respect to physical models, by considering the accuracy and stability of the models and by referring to the other CFD researches [1,2,6] of steam methane reformers, the standard k-ε model [13], discrete ordinate (DO) radiation model [14] and finite rate/eddy dissipation (FRED) model [15] are adopted for turbulence, radiation and chemical reaction simulations, respectively. The standard wall functions [16] are used to resolve the flow quantities (velocity, temperature, and turbulence quantities) at the near-wall regions.

For the steady-state three-dimensional flow field with chemical reaction in this study, the governing equations include the continuity equation, momentum equation, turbulence model equation (k-ε model), energy equation, radiation model equation (discrete ordinate radiation model), and chemical reaction model equation (FRED model). Among these models, only the FRED chemical reaction model is described below while the others and the convergence criterion are not described because they have been introduced in the author's previous study [17].

Consider the general form of the rth chemical reaction as follows:

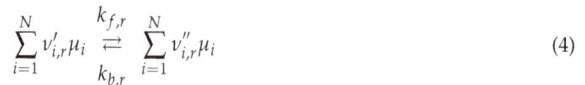

$$\sum_{i=1}^{N} v'_{i,r} \mu_i \underset{k_{b,r}}{\overset{k_{f,r}}{\rightleftharpoons}} \sum_{i=1}^{N} v''_{i,r} \mu_i \tag{4}$$

where

N = number of chemical species in the system
$v'_{i,r}$ = stoichiometric coefficient for reactant i in reaction r
$v''_{i,r}$ = stoichiometric coefficient for product i in reaction r
μ_i = species i
$k_{f,r}$ = forward rate constant for reaction r
$k_{b,r}$ = backward rate constant for reaction r

Equation (4) is valid for both reversible and non-reversible reactions. For non-reversible reactions, the backward rate constant, $k_{b,r}$, is omitted. The species transport equation of a chemical reaction system can be written as

$$\nabla \cdot (\rho \vec{v} Y_i) = \nabla \cdot \left(\rho D_{i,m} + \frac{\mu_t}{Sc_t} \right) \nabla Y_i + R_i + S_i \tag{5}$$

where Y_i, $D_{i,m}$, Sc_t, R_i, and S_i are the mass fraction, diffusion coefficient, turbulent Schmidt number, net generation rate, and extra source term of species i, respectively. The net source of chemical species i due to reaction is computed as the sum of the Arrhenius reaction sources over the N_R reactions that the species participate in $\hat{R}$

$$R_i = M_{w,i} \sum_{r=1}^{N_R} \hat{R}_{i,r} \tag{6}$$

where $M_{w,i}$ is the molecular weight of species i and $\hat{R}_{i,r}$ is the Arrhenius molar rate of creation/destruction of species i in reaction r. For a non-reversible reaction, the molar rate of creation/destruction of species i in reaction r is given by

$$\hat{R}_{i,r} = \Gamma(v_{i,r}'' - v_{i,r}') \left(k_{f,r} \prod_{j=1}^{N} [C_{j,r}]^{(\eta_{j,r}' + \eta_{j,r}'')} \right) \tag{7}$$

For a reversible reaction,

$$\hat{R}_{i,r} = \Gamma(v_{i,r}'' - v_{i,r}') \left(k_{f,r} \prod_{j=1}^{N} [C_{j,r}]^{\eta_{j,r}'} - k_{b,r} \prod_{j=1}^{N} [C_{j,r}]^{v_{j,r}''} \right) \tag{8}$$

where

$C_{j,r}$ = molar concentration of species j in reaction r (kgmol/m^3)
$\eta_{j,r}'$ = rate exponent for reactant species j in reaction r
$\eta_{j,r}''$ = rate exponent for product species j in reaction r
Γ = net effect of third bodies on the reaction rate

The forward and backward rate constants for reaction r, $k_{f,r}$ and $k_{b,r}$, are computed using the Arrhenius expression:

$$k_{f,r} = A_r T^{\beta_r} e^{-E_r/RT} \tag{9}$$

$$k_{b,r} = \frac{k_{f,r}}{K_r} \tag{10}$$

where

A_r = pre-exponential factor (consistent units)
β_r = temperature exponent (dimensionless)
E_r = activation energy for the reaction (J/kgmol)
R = universal gas constant (J/kgmol-K)

K_r is the equilibrium constant for the rth reaction and is computed from

$$K_r = \exp\left(\frac{\Delta S_r^0}{R} - \frac{\Delta H_r^0}{RT} \right) \left(\frac{p_{atm}}{RT} \right)^{\sum_{i=1}^{N} (v_{i,r}'' - v_{i,r}')} \tag{11}$$

where p_{atm} denotes atmospheric pressure (101, 325 Pa). The term within the exponential function represents the change in Gibbs free energy, and its components are computed as follows:

$$\frac{\Delta S_r^0}{R} = \sum_{i=1}^{N} \left(v_{i,r}'' - v_{i,r}' \right) \frac{S_i^0}{R} \tag{12}$$

$$\frac{\Delta H_r^0}{RT} = \sum_{i=1}^{N} \left(v_{i,r}'' - v_{i,r}' \right) \frac{h_i^0}{RT} \tag{13}$$

where S_i^0 and h_i^0 are the standard-state entropy and standard-state enthalpy (heat of formation). In this study, the kinetic and thermodynamic constants for reactions (1)–(3) used in Lemnouer Chibane and Brahim Djellouli's work [18] are adopted.

In general, a reformer operates at high temperatures. For a non-premixed reaction, turbulence mixes the reactants and then advects the mixture to the reaction zone for quick reaction. For a premixed reaction, turbulence mixes the lower-temperature reactants and the higher-temperature products and then advects the mixture to the reaction zone for a quick reaction. Therefore, the chemical reaction is generally mixing (diffusion) controlled. However, the flue gas, fuel gas and air are generally premixed before injecting into the reformer. Although the chemical reaction in most regions in a reformer is mixing controlled, in some regions, e.g., the neighborhood of the feed inlet, the chemical reaction is kinetically controlled. In existing chemical reaction models, the eddy dissipation model (EDM) [19] can consider simultaneously the diffusion controlled and the kinetically controlled reaction rates. In EDM, the net generation rate of species i in the rth chemical reaction is found from the smaller value of the following two reaction rates:

$$R_{i,r} = v'_{i,r} M_{w,i} A \rho \frac{\varepsilon}{k} \min_R \left(\frac{Y_R}{v'_{R,r} M_{w,R}} \right) \tag{14}$$

$$R_{i,r} = v'_{i,r} M_{w,i} A B \rho \frac{\varepsilon}{k} \frac{\sum_P Y_P}{\sum_j^N v''_{j,r} M_{w,j}} \tag{15}$$

where

Y_P is the mass fraction of any product, P

Y_R is the mass fraction of a particular reactant, R

A is an empirical constant equal to 4.0

B is an empirical constant equal to 0.5

In general, the EDM works well for a non-premixed reaction. However, for a premixed reaction, the reaction may start immediately when injecting into a reformer. This is unrealistic in practical situations. To overcome this unreasonable phenomenon, ANSYS FLUENT provides another model, the Finite-Rate/Eddy-Dissipation (FRED) model, which combines the finite-rate model and the EDM. In this model, the net generation rate of a species is taken as the smaller value of the Arrhenius reaction rate and the value determined by EDM. The Arrhenius reaction rate plays the role of a switch to avoid the unreasonable situation that the reaction starts immediately when injecting into a reformer. Once the reaction is activated, the eddy-dissipation rate is generally lower than the Arrhenius reaction rate, and the reaction rate is then determined by the EDM.

3. Results and Discussion

To validate the numerical methods and physical models used in this study, an industrial-scale steam methane reformer is simulated. The configuration and dimension of the reformer investigated is shown in Figure 1. Note that only one half of the reformer is simulated due to its symmetry, as shown in Figure 1b,c. The reformer contains 138 reforming tubes and 216 burners on one side (totally 276 tubes and 432 burners). The outer diameter and thickness of a reforming tube are 136 mm and 13.4 mm, respectively, while the diameter of a burner is 197 mm.

(**a**) a typical steam methane reformer   (**b**) numerical model of the steam methane reformer

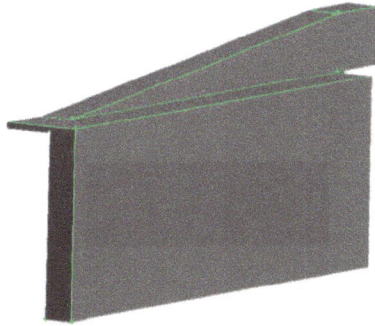

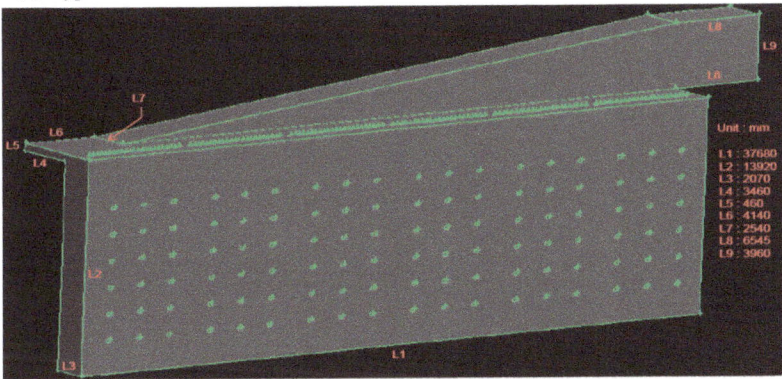

(**c**) dimension of the steam methane reformer

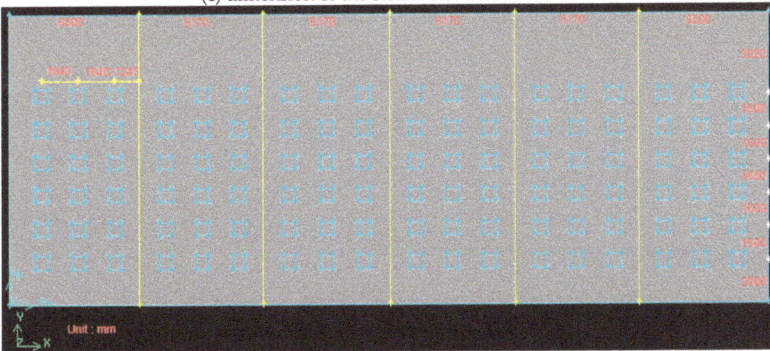

(**d**) illustration of the burner positions

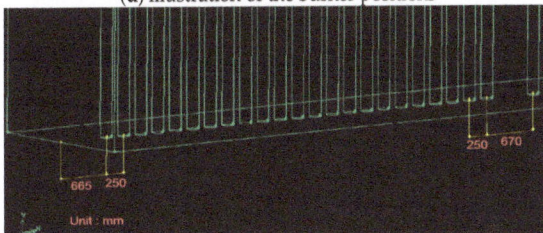

(**e**) illustration of the reforming tube positions

Figure 1. Configuration and dimension of the steam methane reformer investigated.

The boundary conditions for the numerical model of the steam methane reformer are described below. These conditions are practical operating conditions that are used by a petrochemical corporation in Taiwan.

(1) Symmetry plane: symmetric boundary condition
(2) Wall: standard wall function
(3) Reforming tube inlet:

V = 5.4 m/s (in axial direction)

T = 912.75 K

$P_{gauge} = 2.1658 \times 10^6$ N/m^2

Species mole fraction:

$CH_4 = 0.2029$
$H_2O = 0.6$
$H_2 = 0.12855$
$CO_2 = 0.06565$
$CO = 0.00145$
$N_2 = 0.00145$

(4) Reforming tube exit: The diffusion flux for all flow variables in the outflow direction are zero. In addition, the mass conservation is obeyed at the exit.
(5) Fuel and flue gas inlet (burner inlet):

V = 2.404 m/s (in radial direction)

T = 673.15 K

$P_{gauge} = 1.04544 \times 10^4$ N/m^2

Species mole fraction:

$H_2 = 0.0816$
$CH_4 = 0.0474$
$N_2 = 0.49057$
$O_2 = 0.12818$
$CO_2 = 0.25225$

(6) Fuel and flue gas exit: The diffusion flux for all flow variables in the outflow direction are zero. In addition, the mass conservation is obeyed at the exit.

The turbulence kinetic energy is 10% of the inlet mean flow kinetic energy and the turbulence dissipation rate is computed using Equation (16).

$$\varepsilon = C_\mu^{3/4} \frac{k^{3/2}}{l} \tag{16}$$

where $l = 0.07$ L and L is the hydraulic diameter.

3.1. Comparison of Numerical Results with Experimental Data

The simulation results are compared with the experimental data from a petrochemical refinery in Taiwan to evaluate the numerical methods and physical models adopted in this study. As mentioned above, the real reformer contains 138 reforming tubes and 216 burners on one side (totally 276 tubes and 432 burners). To save simulation time, a simplified model is also calculated and compared. The simplified model contains 6 tubes and 12 burners on one side of the reformer. The arrangement of reforming tubes and burners as well as their dimensions for the simplified model are shown in

Figure 2. The flowrates in the reforming tubes and burners for the simplified model are the same as those in the real reformer. Therefore, the reforming tubes and burners for the simplified model have larger diameters.

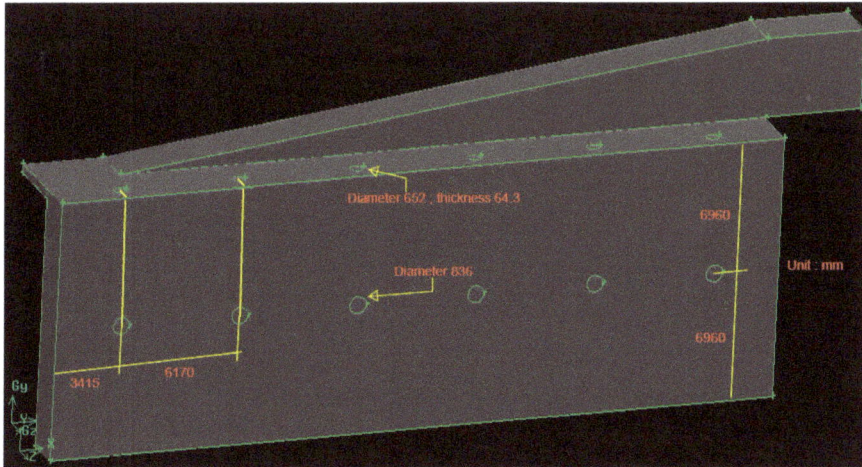

Figure 2. The arrangement of reforming tubes and burners as well as their dimensions for the simplified model.

The numbers of CFD cells for the real reformer model and the simplified model are around 4 million and 1.5 million, respectively. The grid mesh is generated by the software GAMBIT and is unstructured. The dimensionless distance from the wall, y^*, in the wall function method has been examined after a converged solution is obtained. It was found that the values of y^* for the nodes at the wall to their nearest interior nodes vary between 20.0 and 60.0, and lie in the logarithmic layer of the wall function method. This implies that the wall-adjacent cells of the grid mesh in this study are suitable for the use of wall function.

The computer used in this study is an ASUS ESC-500-G4 work station of 8 cores with Intel Core i7-6700 CPU and 64 GB ram. The solution of the CFD model for the real reformer is obtained after approximately 30 full days while that for the simplified reformer is approximately 10 full days.

Figure 3 compares the average temperatures at the outer surfaces of the reforming tubes. It can be seen that the simulation result agrees well with the experimental data. The deviations from the experimental data using the real reformer model and the simplified model are 2.88% and 3.18%, respectively, which are both acceptable from a viewpoint of engineering applications. The result calculated from the real reformer model agrees better with the experimental data than that from the simplified model, although the latter also gives an acceptable result.

In the subsequent discussion, the simplified model is used for parametric study to save simulation time.

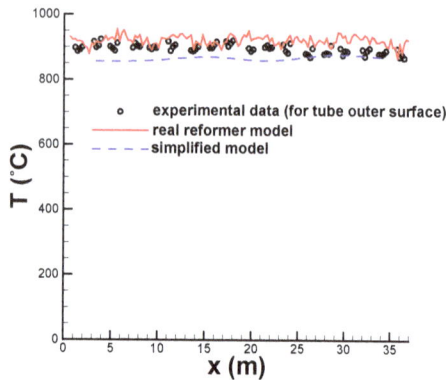

Figure 3. Comparison of the average temperatures at the outer surfaces of the reforming tubes.

3.2. Effect of the Mole Fraction of Steam

To explore the effect of steam quantities used in the SMR process on the hydrogen yield, seven different mole fractions of steam (Y_{H2O}), 0.1, 0.2, 0.3, 0.4, 0.5, 0.6, and 0.7 are calculated and compared. The mole fraction of steam for practical operation is 0.6. The mole fractions of the other species in the reforming reactants, except methane, are kept unchanged. This implies that the mole fraction of methane decreases when the mole fraction of steam increases.

Figure 4 compares the average temperatures at the surfaces of the reforming tubes using different mole fractions of steam. From the simulation results, the lowest average temperatures at the surfaces (outer or inner) of the reforming tubes occur when Y_{H2O} is 0.5. The reforming reactants with a steam mole fraction of 0.5 are approximately a stoichiometric mixture, which has a more complete reforming reaction. From Figure 4, it is also observed that the inner surface temperature of the reforming tube is lower than its outer surface temperature at this inlet velocity of reforming reactants (5.4 m/s). The effect of inlet velocities of the reforming reactants will be discussed in the next section.

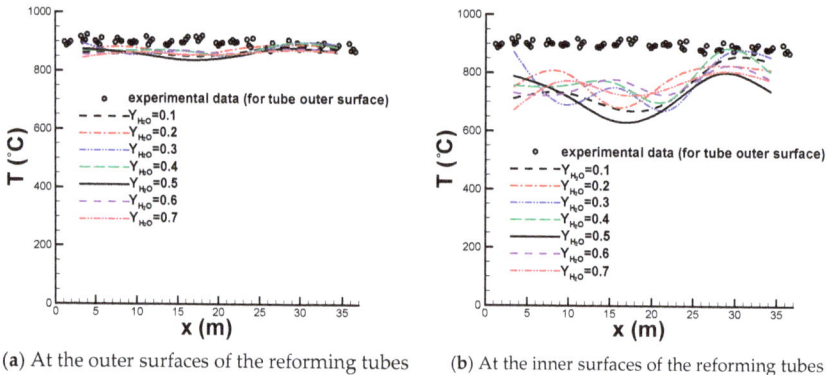

(**a**) At the outer surfaces of the reforming tubes   (**b**) At the inner surfaces of the reforming tubes

Figure 4. Comparison of the average temperatures at the surfaces of the reforming tubes using different mole fractions of steam.

The above result can also be observed from Figure 5 which compares the average mole fractions of hydrogen at the reforming tube outlets using different mole fractions of steam. As expected, the hydrogen yield has the highest value when Y_{H2O} is 0.5 because of a more complete reforming reaction. When Y_{H2O} is less than 0.5, the hydrogen yield increases with Y_{H2O}. On the other hand, when Y_{H2O} is greater than 0.5, the hydrogen yield decreases with Y_{H2O}. Further, it can be found from Figure 5 that the simulated hydrogen yield is 0.708 for the steam mole fraction of real operation (i.e., $Y_{H2O} = 0.6$). The real value of the hydrogen yield is 0.698. The deviation of the CFD simulation is 1.43%

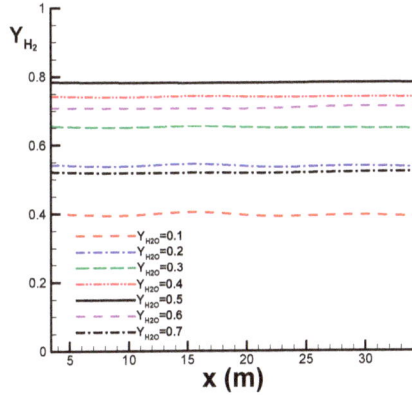

Figure 5. Comparison of the average mole fractions of hydrogen at the reforming tube outlets using different mole fractions of steam.

When a steam reformer is operating, the catalyst tubes are subjected to stresses close to the ultimate stress of the tube material. This leads to an acceleration of the creep damage. Figure 6 compares the average wall shear stresses at the surfaces of the reforming tubes using different mole fractions of steam. It can be observed that higher wall shear stresses at the surfaces of the reforming tubes occur at higher Y_{H2O}. For example, the highest wall shear stress at the surfaces of the reforming tubes occurs when Y_{H2O} is 0.7. From Figure 6, it is also observed that the wall shear stresses at the inner surfaces of the reforming tubes are higher than those at their outer surfaces.

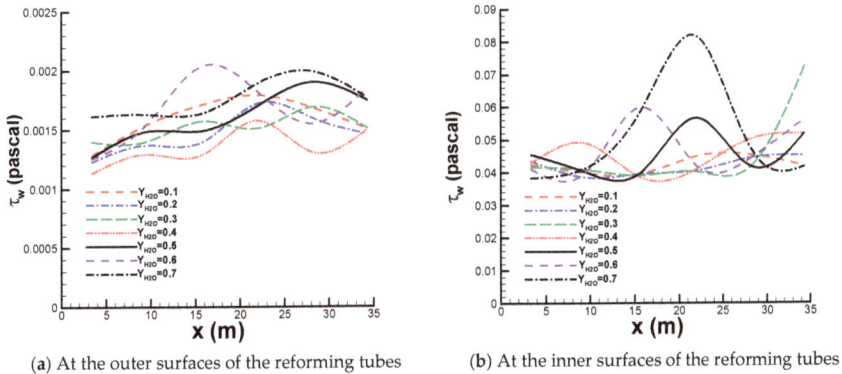

(**a**) At the outer surfaces of the reforming tubes (**b**) At the inner surfaces of the reforming tubes

Figure 6. Comparison of the average wall shear stresses at the surfaces of the reforming tubes using different mole fractions of steam.

To explore the influence of the flow and thermal development on the wall shear stress, Figure 7 compares the temperature and velocity vector distributions at $y = 6$ m around a reforming tube using different mole fractions of steam. It is observed that the wall shear stress is closely connected with the velocity vector distributions around the reforming tube. A faster flow through a reforming tube results in a higher wall shear stress.

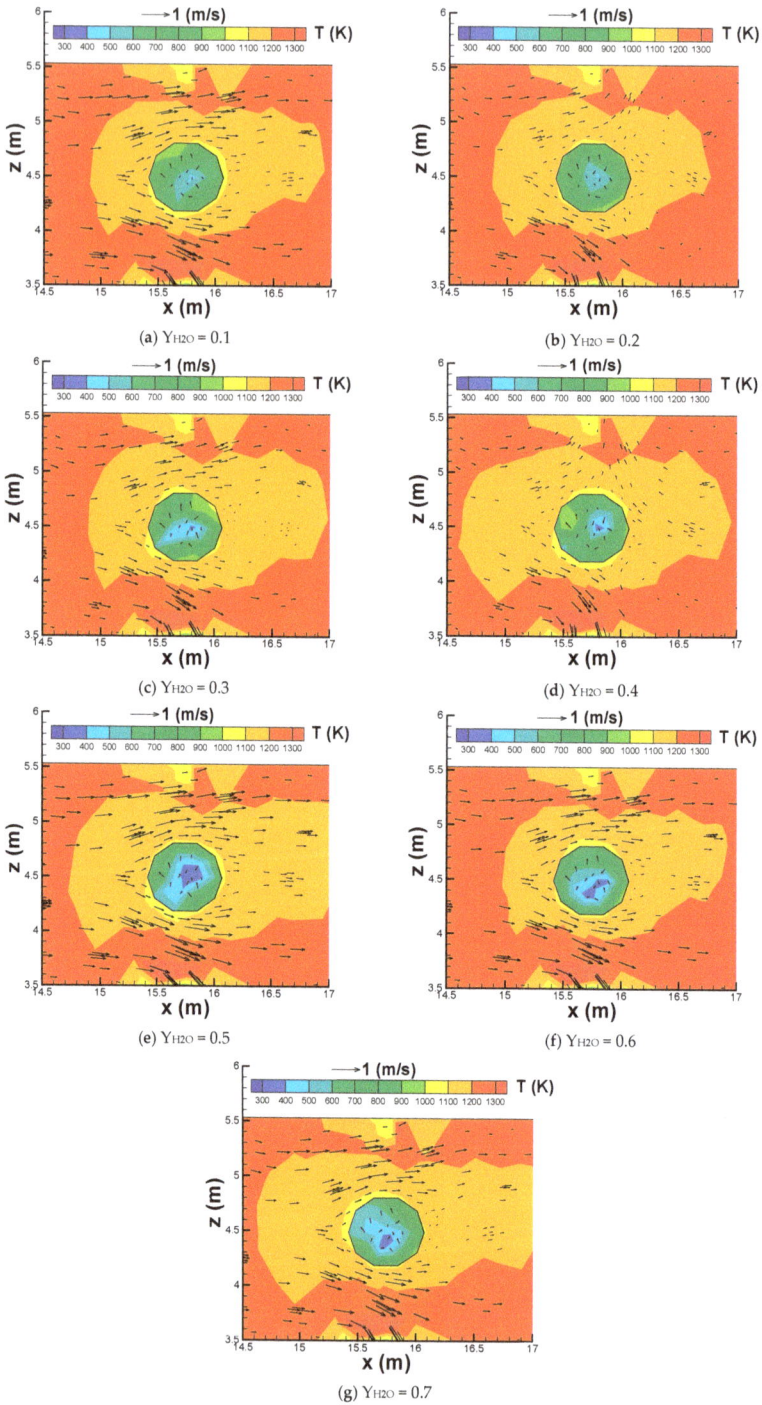

(a) $Y_{H2O} = 0.1$

(b) $Y_{H2O} = 0.2$

(c) $Y_{H2O} = 0.3$

(d) $Y_{H2O} = 0.4$

(e) $Y_{H2O} = 0.5$

(f) $Y_{H2O} = 0.6$

(g) $Y_{H2O} = 0.7$

Figure 7. Comparison of the temperature and velocity vector distributions at y = 6 m around a reforming tube using different mole fractions of steam.

3.3. Effect of Inlet Velocities of the Reforming Reactants

To explore the effect of inlet velocities of the reforming reactants on hydrogen yield, five different inlet velocities of the reforming reactants ($V_{reforming\ reactants}$), 1.35, 2.7, 5.4, 8.1 and 10.8 m/s are calculated and compared. The inlet velocity of reforming reactants for the real operation is 5.4 m/s. The mole fraction of steam in the reforming reactants is taken as 0.5 because it is close to a stoichiometric mixture, which has a more complete reforming reaction. The other operating conditions, including pressure, temperature, velocity, and species composition, on the burner side and reforming tube side are kept unchanged.

Figure 8 compares the average temperatures at the surfaces of the reforming tubes using different inlet velocities of the reforming reactants. It is observed that the surface temperatures (outer and inner) of the reforming tubes increase with the inlet velocities of the reforming reactants. It should be noted that the tolerance of temperature for a reforming tube is around 950 °C. Therefore, the inlet velocity of the reforming reactants should not exceed the real value 5.4 m/s too much. When the inlet velocities of the reforming reactants are lower, the outer surface temperatures of the reforming tubes are higher than their inner surface temperatures. On the other hand, when the inlet velocities of the reforming reactants are higher, the inner surface temperatures of the reforming tubes are higher than their outer surface temperatures. From Figure 8, it is also observed that the inner surface temperatures of the reforming tubes are more sensitive to the inlet velocities of the reforming reactants, as compared to their outer surface temperatures.

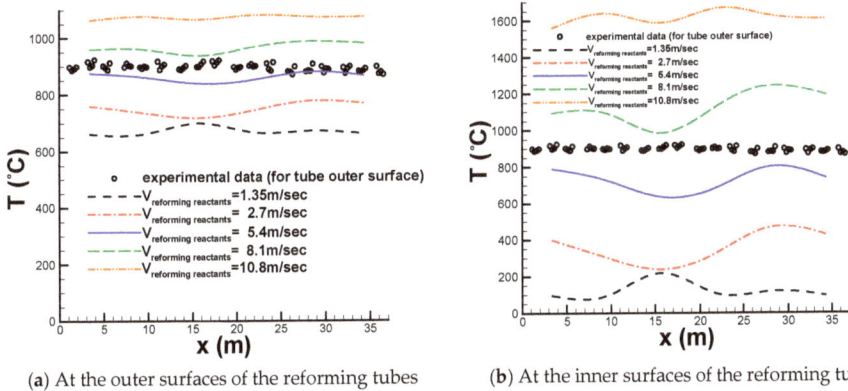

(a) At the outer surfaces of the reforming tubes

(b) At the inner surfaces of the reforming tubes

Figure 8. Comparison of the average temperatures at the surfaces of the reforming tubes using different inlet velocities of the reforming reactants.

Figure 9 compares the average mole fractions of hydrogen at the reforming tube outlets using different inlet velocities of the reforming reactants. Because the mole fractions of the reforming reactants are kept unchanged, it is observed that the inlet velocity of the reforming reactants has a minor influence on the hydrogen yields at the reforming tube outlets. However, because the flowrate of the reforming reactants is proportional to their inlet velocity, the hydrogen yields are naturally increased when the inlet velocity of the reforming reactants is increased.

Figure 10 compares the average wall shear stresses at the surfaces of the reforming tubes using different inlet velocities of the reforming reactants. It is observed that the wall shear stresses at the surfaces of the reforming tubes increase with the inlet velocity of the reforming reactants. Similar to the results using different mole fractions of steam, the wall shear stresses at the inner surfaces of the reforming tubes are higher than those at their outer surfaces. To explore the influence of the flow and thermal development on the wall shear stress, Figure 11 shows the comparison of the temperature and velocity vector distributions at y = 6 m around a reforming tube using different inlet velocities of the reforming reactants. Similar to the results using different mole fractions of steam, it is found that the wall shear stress is closely connected with the velocity vector distributions around a reforming tube. A faster fluid flow through a reforming tube results in a larger wall shear stress. From Figure 11 it is also observed that the temperature inside and outside a reforming tube increases with the inlet velocity of the reforming reactants. The overall reaction inside a reforming tube is endothermic. Increasing the velocity of the reforming reactants also increases the reforming tube temperatures due to the higher heat absorption of the reforming reactants. This also increases the temperature outside the reforming tubes due to the radiation emission from the reforming tubes. The tolerance of the reforming tube temperature, 950 °C, may be exceeded. Therefore, it should be careful when using higher velocities of the reforming reactants to increase the hydrogen production. The inlet velocity of the reforming reactants should not exceed the real value (5.4 m/s) too much.

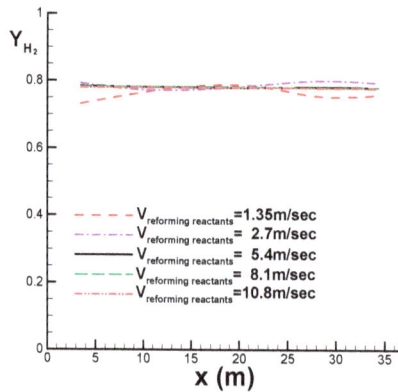

Figure 9. Comparison of the average mole fractions of hydrogen at the reforming tube outlets using different inlet velocities of the reforming reactants.

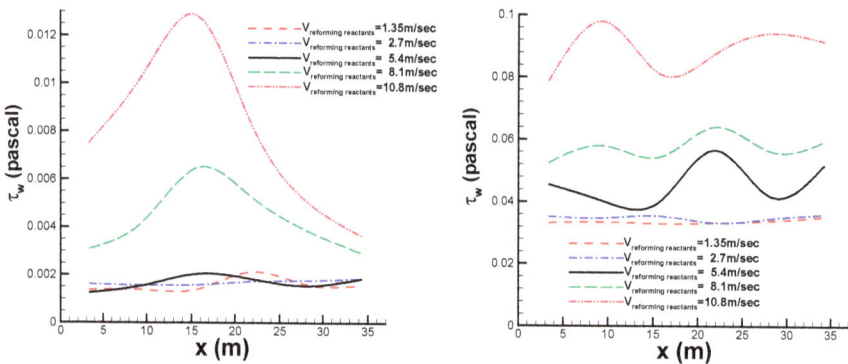

(a) At the outer surfaces of the reforming tubes   (b) At the inner surfaces of the reforming tubes

Figure 10. Comparison of the average wall shear stresses at the surfaces of the reforming tubes using different inlet velocities of the reforming reactants.

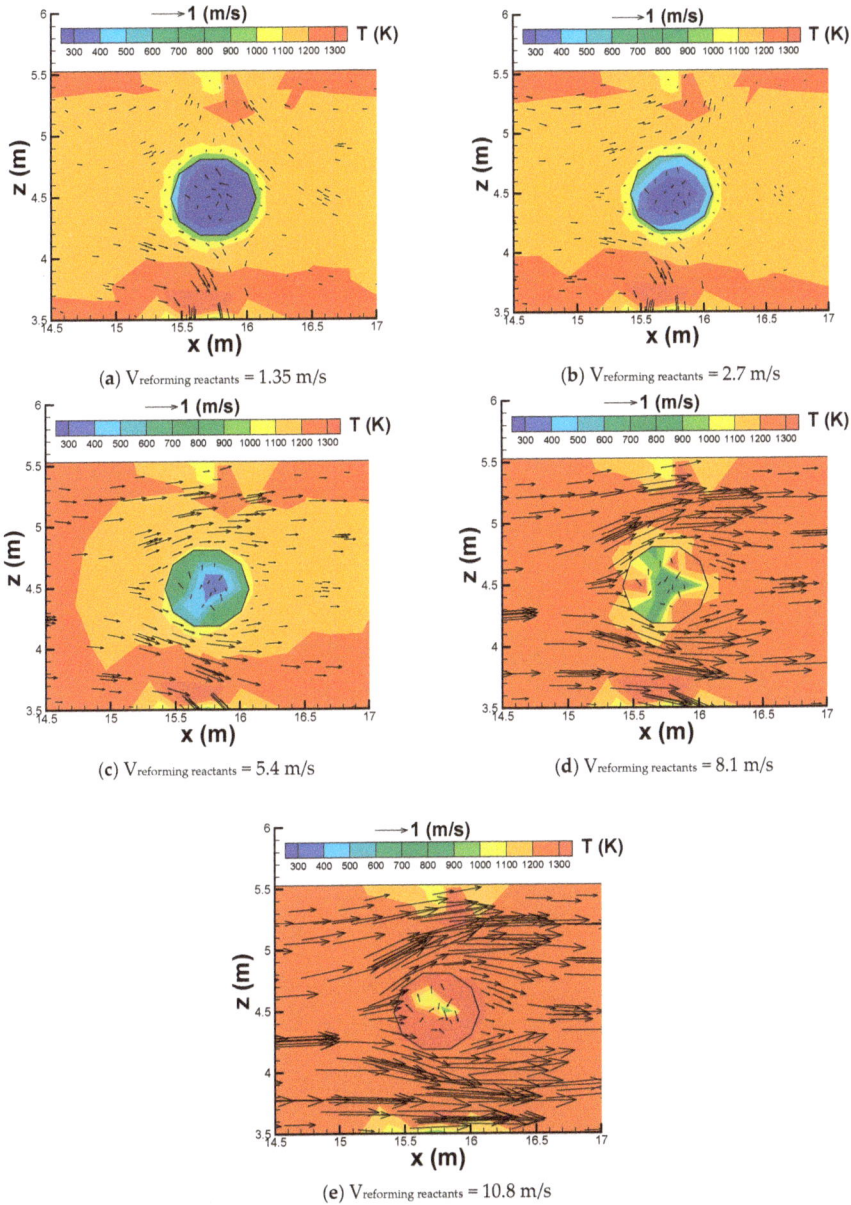

(**a**) $V_{\text{reforming reactants}} = 1.35$ m/s

(**b**) $V_{\text{reforming reactants}} = 2.7$ m/s

(**c**) $V_{\text{reforming reactants}} = 5.4$ m/s

(**d**) $V_{\text{reforming reactants}} = 8.1$ m/s

(**e**) $V_{\text{reforming reactants}} = 10.8$ m/s

Figure 11. Comparison of the temperature and velocity vector distributions at y = 6 m around a reforming tube using different inlet velocities of the reforming reactants.

4. Conclusions

In this paper, the transport and chemical reaction in an industrial-scale steam methane reformer are simulated using CFD. Two factors influencing the reformer temperature, hydrogen yield and stress distribution are discussed: (1) the mole fraction of steam and (2) the inlet velocity of reforming reactants. The purpose of this paper is to get a better understanding of the flow and thermal development in a steam reformer and thus, to make it possible to improve the performance and lifetime of a steam reformer. From the simulation results, it is found that the lowest temperature at the reforming tube surfaces occurs when Y_{H2O} is 0.5. Hydrogen yield has the highest value when Y_{H2O} is 0.5. Hydrogen yield increases with Y_{H2O} when Y_{H2O} is less than 0.5 and decreases with Y_{H2O} when Y_{H2O} is greater than 0.5. The wall shear stress at a reforming tube surface is higher at a higher Y_{H2O}. A faster fluid flow through a reforming tube results in a larger wall shear stress. The surface temperature of a reforming tube increases when the inlet velocity of the reforming reactants increases. When the inlet velocity of the reforming reactants is lower, the outer surface temperature of a reforming tube is higher than its inner surface temperature. On the other hand, when the inlet velocity of the reforming reactants is higher, the inner surface temperature of a reforming tube is higher than its outer surface temperature. As compared to the outer surface temperature of a reforming tube, its inner surface temperature is more sensitive to the inlet velocity of the reforming reactants. Finally, the wall shear stress at a reforming tube surface increases when the inlet velocity of the reforming reactants increases.

Funding: This research and the APC were funded by the Ministry of Science and Technology, Taiwan, under the contract MOST106-2221-E-150-061.

Acknowledgments: The author is grateful to the Formosa Petrochemical Corporation in Taiwan for providing valuable data and constructive suggestions to this research during the execution of the industry-university cooperative research project under the contract 104AF-86.

Conflicts of Interest: The author declares no conflict of interest.

Abbreviations

The following abbreviations are used in this manuscript:

$C\mu$	turbulence model constant (=0.09)
K	turbulence kinetic energy (m^2/s^2)
P	pressure (N/m^2)
T	temperature (K)
V	velocity (m/s)
XYZ	cartesian coordinates with origin at the centroid of the burner inlet (m)
Y	mole fraction (%)
Greek symbols	
ε	turbulence dissipation rate (m^2/s^3)
μ	viscosity (kg/(m s))
ρ	density (kg/m^3)
τ	shear stress (N/m^2)

References

1. Tran, A.; Aguirre, A.; Durand, H.; Crose, M.; Christofides, P.D. CFD modeling of an industrial-scale steam methane reforming furnace. *Chem. Eng. Sci.* **2017**, *171*, 576–598. [CrossRef]
2. Di Carlo, A.; Aloisi, I.; Jand, N.; Stendardo, S.; Foscolo, P.U. Sorption enhanced steam methane reforming on catalyst-sorbent bifunctional particles: A CFD fluidized bed reactor model. *Chem. Eng. Sci.* **2017**, *173*, 428–442. [CrossRef]
3. Irani, M.; Alizadehdakhel, A.; Nakhaei Pour, A.; Hoseini, N.; Adinehnia, M. CFD modeling of hydrogen production using steam reforming of methane in monolith reactors: Surface or volume-base reaction model? *Int. J. Hydrogen Energy* **2011**, *36*, 15602–15610. [CrossRef]
4. Kimmel, A.S. Heat and Mass Transfer Correlations for Steam Methane Reforming in Non-Adiabatic, Process-Intensified Catalytic Reactors. Master's Thesis, Marquette University, Milwaukee, WI, USA, 2011.
5. Castagnola, L.; Lomonaco, G.; Marotta, R. Nuclear systems for hydrogen production: State of art and perspectives in transport sector. *Glob. J. Energy Technol. Res. Updat.* **2014**, *1*, 4–18.
6. Lao, L.; Aguirre, A.; Tran, A.; Wu, Z.; Durand, H.; Christofides, P.D. CFD modeling and control of a steam methane reforming reactor. *Chem. Eng. Sci.* **2016**, *148*, 78–92. [CrossRef]

7. Mokheimer, E.M.A.; Hussain, M.I.; Ahmed, S.; Habib, M.A.; Al-Qutub, A.A. On the modeling of steam methane reforming. *J. Energy Resour. Technol.* **2015**, *137*, 012001. [CrossRef]

8. Ni, M. 2D heat and mass transfer modeling of methane steam reforming for hydrogen production in a compact reformer. *Energy Convers. Manag.* **2013**, *65*, 155–163. [CrossRef]

9. Ebrahimi, H.; Soltan Mohammadzadeh, J.S.; Zamaniyan, A.; Shayegh, F. Effect of design parameters on performance of a top fired natural gas reformer. *Appl. Therm. Eng.* **2008**, *28*, 2203–2211. [CrossRef]

10. Seo, Y.-S.; Seo, D.-J.; Seo, Y.-T.; Yoon, W.-L. Investigation of the characteristics of a compact steam reformer integrated with a water-gas shift reactor. *J. Power Sources* **2006**, *161*, 1208–1216. [CrossRef]

11. Fluent Inc. *ANSYS FLUENT 17 User's Guide*; Fluent Inc.: New York, NY, USA, 2017.

12. Patankar, S.V. *Numerical Heat Transfer and Fluid Flows*; McGraw-Hill: New York, NY, USA, 1980.

13. Launder, B.E.; Spalding, D.B. *Lectures in Mathematical Models of Turbulence*; Academic Press: London, UK, 1972.

14. Siegel, R.; Howell, J.R. *Thermal Radiation Heat Transfer*; Hemisphere Publishing Corporation: Washington, DC, USA, 1992.

15. Sivathanu, Y.R.; Faeth, G.M. Generalized state relationships for scalar properties in non-premixed hydrocarbon/air flames. *Combust. Flame* **1990**, *82*, 211–230. [CrossRef]

16. Launder, B.E.; Spalding, D.B. The numerical computation of turbulent flows. *Comput. Method Appl. Mech.* **1974**, *3*, 269–289. [CrossRef]

17. Yeh, C.L. Numerical analysis of the combustion and fluid flow in a carbon monoxide boiler. *Int. J. Heat Mass Transf.* **2013**, *59*, 172–190. [CrossRef]

18. Chibane, L.; Djellouli, B. Methane steam reforming reaction behaviour in a packed bed membrane reactor. *Int. J. Chem. Eng. Appl.* **2011**, *2*, 147–156. [CrossRef]

19. Magnussen, B.F.; Hjertager, B.H. On mathematical models of turbulent combustion with special emphasis on soot formation and combustion. In *Symposium (international) on Combustion*; The Combustion Institute: Pittsburgh, PA, USA, 1976.

MDPI

St. Alban-Anlage 66

4052 Basel

Switzerland

Tel. +41 61 683 77 34

Fax +41 61 302 89 18

www.mdpi.com

Energies Editorial Office

E-mail: energies@mdpi.com

www.mdpi.com/journal/energies